About Island Press

Island Press is the only nonprofit organization in the United States whose principal purpose is the publication of books on environmental issues and natural resource management. We provide solutions-oriented information to professionals, public officials, business and community leaders, and concerned citizens who are shaping responses to environmental problems.

In 2003, Island Press celebrates its nineteenth anniversary as the leading provider of timely and practical books that take a multidisciplinary approach to critical environmental concerns. Our growing list of titles reflects our commitment to bringing the best of an expanding body of literature to the environmental community throughout North America and the world.

Support for Island Press is provided by The Nathan Cummings Foundation, Geraldine R. Dodge Foundation, Doris Duke Charitable Foundation, Educational Foundation of America, The Charles Engelhard Foundation, The Ford Foundation, The George Gund Foundation, The Vira I. Heinz Endowment, The William and Flora Hewlett Foundation, Henry Luce Foundation, The John D. and Catherine T. MacArthur Foundation, The Andrew W. Mellon Foundation, The Moriah Fund, The Curtis and Edith Munson Foundation, National Fish and Wildlife Foundation, The New-Land Foundation, Oak Foundation, The Overbrook Foundation, The David and Lucile Packard Foundation, The Pew Charitable Trusts, The Rockefeller Foundation, The Winslow Foundation, and other generous donors.

The opinions expressed in this book are those of the author(s) and do not necessarily reflect the views of these foundations.

About the Sea Around Us Project

The Sea Around Us Project is a partnership of the University of British Columbia Fisheries Centre and the Pew Charitable Trusts. The aims of the project are to provide an integrated analysis of the impacts of fisheries on marine ecosystems, and to devise policies that can mitigate and reverse harmful trends while ensuring the social and economic benefits of sustainable fisheries. The project started in July 1999 and is led by Daniel Pauly. The project's name is taken from Rachel Carson's *The Sea Around Us*, published in 1951.

IN A PERFECT OCEAN

THE STATE OF THE WORLD'S OCEANS SERIES
Daniel Pauly, series editor

In a Perfect Ocean: The State of Fisheries and Ecosystems in the North Atlantic Ocean, by Daniel Pauly and Jay Maclean

IN A PERFECT OCEAN

THE STATE OF FISHERIES AND ECOSYSTEMS IN THE NORTH ATLANTIC OCEAN

DANIEL PAULY
AND
JAY MACLEAN

The Sea Around Us Project

ISLAND PRESS

Washington • Covelo • London

©2003 Island Press

Library of Congress Cataloging-in-Publication Data

Pauly, D. (Daniel)
   In a perfect ocean : the state of fisheries and ecosystems in the North Atlantic Ocean / Daniel Pauly and Jay Maclean.
      p. cm. — (The state of the world's oceans series)
Includes bibliographical references (p.   ).
   ISBN 1-55963-323-9 (hardcover : alk. paper) — ISBN 1-55963-324-7 (pbk. : alk. paper)
   1. Fisheries—North Atlantic Ocean. 2. Marine ecology—North Atlantic Ocean. I. Maclean, J. L. (Jay L.) II. Title. III. Series.
   SH213.2 .P38 2002
333.95'6'091631—dc21
                                                                            2002152291

British Cataloguing-in-Publication Data available.

Printed on recycled, acid-free paper ✪

Design by Artech Group, Inc.

Manufactured in the United States of America
10 9 8 7 6 5 4 3 2 1

To Margie and Marlon, and to Sandra, Ilya, and Angela

Contents

List of Figures and Maps

food web, that is, from large finfishes (especially cod) to smaller finfishes and to invertebrates, especially mollusks such as clams.

Fishing down marine food webs mean that the fisheries, having at first removed the larger fishes at the top of various food chains, must target fishes lower and lower down, and end up targeting very small fishes and plankton, including jellyfish.

Less enduring than their namesakes in Egypt, the ocean's food webs, which can be conveniently represented as pyramids, have been "squashed" by a century of unsustainable fishing.

Overlap between the prey of marine mammals and the catch of fisheries in the North Atlantic in the 1990s.

Breakdown of the estimated 2.5 billion US$ in annual fisheries subsidies, by country/region of the North Atlantic, and by type of subsidy.

Comparisons of the small- vs. large-scale subsectors in Norwegian fisheries, using data for 1998.

Parts of the North Atlantic covered by various international instruments devoted to fisheries management or environmental protection.

The major types of fish under the care of the six international instruments are all in a state of decline.

Preface

This volume, the first in a series, presents the findings of an ambitious project—to measure the impact of fishing on the ecosystems that make up the North Atlantic Ocean and to propose ways to mitigate that impact. The project arose from a request by Dr. Joshua Reichert, the Director of the Environment Program of the Pew Charitable Trusts, Philadelphia, to answer six specific questions about the North Atlantic:

- What are the total fisheries catches from the ecosystems, including reported and unreported landings and discards at sea?
- What are the biological impacts of these withdrawals of biomass for the remaining life in the ecosystems?
- What would be the likely biological and economic impacts of continuing current fishing trends?
- What were the former states of these ecosystems before the expansion of large-scale commercial fisheries?
- How do the present ecosystems rate on a scale from "healthy" to "unhealthy"?
- What specific policy changes and management measures should be implemented to avoid continued worsening of the present situation and improve the North Atlantic ecosystem's "health"?

These questions were prompted by previous research, which strongly suggested that fisheries in the North Atlantic and in many other areas are gradually destroying the ecosystems on which they depend. This is an alarming prospect, not only for the fishers and consumers of the ocean's living resources, but for the conservation of its biodiversity, and for the non-extractive uses made of the resources, such as diving, whale- or bird-watching and other forms of ecotourism.

Overfishing—catching too many fish in a given time and area and resulting in a dearth of fish in subsequent years—has been a concern for at least several centuries. Fisheries science was begun in the early twentieth century to advise managers of ways to maximize "sustainable catches." It has largely failed in this endeavor, primarily because the advice it provided went unheeded. Today, though, with the recent shift in attitude toward marine resources as the responsibility of all humankind, not just of a small group of fishers, conservation-oriented scientists are putting forward the case for new arrangements in the stewardship of marine resources.

The project has drawn on fisheries and conservation literature, and has conducted a number of new studies as well, in most cases with new methodologies developed to best answer the questions posed in assessing the many fisheries and ecosystems of the North Atlantic Ocean.

This book offers a comprehensive assessment of fishery impact in the North Atlantic Ocean and recommendations for mitigating that impact. It serves as a model of tested methodologies for analyzing and assessing the condition of other seas and ecosystems as well.

The project was called *The Sea Around Us,* a name drawn from the outstanding book of this title by Rachel Carson.[1] We thank University of British Columbia President, Dr. Martha Piper, for reminding us of that work, and thus inspiring the name of our project.

We hope that through this book, readers will realize the importance of maintaining and safeguarding marine ecosystems, which are in many ways as indispensable to our well-being as the terrestrial ecosystems that we inhabit.

Daniel Pauly, Vancouver
Jay Maclean, Manila

Acknowledgments

We would like to acknowledge the Environment Program of the Pew Charitable Trusts and particularly its Director, Dr. J. Reichert, for the farsightedness and initiative shown in sponsoring this work, and hope that the contents of this volume satisfactorily answer the questions that formed the basis of this project—the questions put by Dr. Reichert (see Preface).

We also wish to thank the members of the *Sea Around Us* Project team, mainly at the Fisheries Centre of the University of British Columbia, whose dedication is evident in the outstanding results of the project's investigations. They are (in alphabetical order): Jackie Alder, Villy Christensen, Sylvie Guénette, Nigel Haggan, Gordon Munro, Tony Pitcher, Peter Tyedmers, Ussif Rashid Sumaila, Reg Watson, and Dirk Zeller. They worked with a large group of scientific colleagues and consultants from countries all around the North Atlantic, to whom we also express our thanks.

Several other persons took part in developing methodologies and conclusions included in this book. They include, from the Fisheries Centre, Eny Buchary, Ratana Chuenpagdee, Bridget Ferriss, Felimon Gayanilo Jr., Ahmed Gelchu, Kristin Kaschner, David Preikshot, Lore Ruttan, and Carl Walters. Other scientists who took part included

Alan Longhurst, Jean-Jacques Maguire, Trevor Platt, and Kenneth Sherman. The methodologies were reviewed at a week-long workshop in May 2000 by a group of experts external to the Fisheries Centre: Lee Alverson, Kevern Cochrane, Poul Degnbol, Paul Fanning, and Richard Grainger.

Other external reviewers kindly provided extensive written comments on the proposed methodologies for the methodology-review workshop: Ragnar Arnason, Trond Bjorndal, John Blaxter, Tony Charles, Cutler Cleveland, Michael Fogarty, Ken Frank, Quentin Grafton, Norman Hall, Rögnvaldur Hannesson, Paul Hart, Simon Levin, Pamela Mace, Paul Medley, Lief Nøttestad, David Pimentel, David Ramm, and Saul Saila.

We are extremely grateful for feedback received at a second workshop, in May 2001, by our invited experts David Allison, Nancy Baron, Philip Clapp, Kevern Cochrane, Paul Fanning, Richard Grainger, Jay Nelson, Andy Rosenberg, Carl Safina, and Lisa Speers.

Thanks also to Amy Poon and Yvette Rizzo for reporting during the workshops, and to Gunna Weingartner and Claire Brignall for organizing them.

Thanks to graphic artist Diana MacPhail for improving our graphs, to Mike Weber for helpful comments, and to Todd Baldwin and his group at Island Press for turning our files into a presentable book.

Daniel Pauly, Vancouver
Jay Maclean, Manila

Introduction

The North Atlantic Ocean has always been portrayed as a dangerous, untamed place, a maelstrom of icebergs, sea creatures, and "nor'easters," a place only the bravest, strongest—or most desperate—fishers dared to venture. For those who did, it held vast riches of cod and swordfish, giant bluefin tuna, right whales, and winged skates the size of barn doors. Its turbulence warded off the incursions of fishers, and even in recent decades, it has continued to claim their lives, including those on board the *Andrea Gail*, the swordfishing vessel whose plight was chronicled in the major motion picture, *The Perfect Storm*.

It is a curiously underappreciated fact that the *Andrea Gail* had been at sea a full 38 days—six days' travel from her home port to a remote part of the North Atlantic—when she ran into the convergence of storms that ultimately sank her. She had gone to the very limit of her fuel-oil tether to find swordfish in numbers large enough to make the trip worth the investment. The nearby fishing grounds on Georges Bank were nearly empty, about to be closed to fishing altogether. When looked at from this perspective, the plight of the *Andrea Gail* points to quite a different picture of the North Atlantic, one of a conquered ocean whose vast fish reserves are depleted below reasonable commercial viability. Far from being the scourge of fishing vessels, it is the North Atlantic that suffers.

Far below the surface, along the continental shelf off the coast of Maine where the *Andrea Gail* might once have fished, the bottom is smooth. In many places, except for the tracks left by a few thin worms, the animals have largely left without a trace. The muddy water moves lazily back and forth, as if swinging with the muffled sounds of the waves, hundreds of feet above. There are no fish. But humans have made their mark: beer cans roll back and forth over the bottom, either from a fishing vessel or from one of the cruise ships that ply that part of the ocean. A few feet farther into the murk, we might see torn bits of coarse nylon mesh form a ghostly shape.

Like most of Georges Bank, this area has been trawled. The tracks cannot be missed. They are left by the rollers of a deep bottom trawl, a contraption about the size of a football field, dragged over the ground to catch fish and whatever else lies in its path. Trawlers plough this part of the ocean several times a year. A few years ago, there might have been a reef there, shimmering with small colorful fish darting about beds of gorgonians and other magic, plant-like animals. Now, it is a low, scarred hill.

The trawlers would not have destroyed the reef in one pass. First, they dragged around it, lifting and removing the large boulders that had protected it, like the outer wall of a castle. Thanks to their precise geo-positioning systems, they were able to return day after day and year after year to the exact same place, gradually eroding the outer parts of the reef. Finally, they reached its central core, where the last fish had found refuge.

About a hundred feet up is where most of the fish congregate, where the longlines trail from boats like the *Andrea Gail*. Well off of Georges Bank, you might find tuna, warm-blooded and swift as bullets until caught, now big chunks of cold flesh. They are the target fish, sought because of the huge prices they command in international markets. If they take the bait, they will be turned into sublime dishes such as sashimi or sushi, or steaks for backyard grills—or into cat food, if the fishers retrieve them too late, after their flesh has lost that special flavor.

The longlines also snag swordfish, large shiny knights without armor, whose large eyes, and the warm brain behind them, help them spot prey at depth, but not tell them from baited hooks. Each year, fishers now struggle mightily to land fewer and smaller swordfish, once numbering in the millions in the North Atlantic. Their slender bodies are sliced up for trendy restaurants.

The other fish dangling from the lines were not targeted; they are the "by-catch." Among them are small and large sharks. Some have long pectoral fins, like bird wings; some have long tail fins, like giant underwater squirrels. Once, sharks were rejected when caught by long-liners, but now their fins are cut off to supply a huge million-dollar market for shark-fin soup. The finned, bleeding carcasses are thrown back overboard—thousands of them, day in, day out. A few more decades of long-lining will resolve the problem, as extinct sharks cannot be finned and discarded. A substitute for shark-fin soup will be found: we are an ingenious lot. But the sharks themselves can never be replaced.

Closer to the surface is another set of shapes, jumbled, some pointing up, some down, some dying while still trying to move forward, but all held by the meshes and folds of a drift net. This is an old net that lost its surface marker buoys and has been drifting for a long time—a murderous Flying Dutchman trapping everything in its twenty-mile-wide path. Some of the entangled fish are tuna, which will not be counted toward reported catch quotas. Others are sailfish, the beautiful cousins of the becalmed swordfish. Still others have strange shapes few humans have ever seen. These fish are so rare that museum curators would consider specimens the highlight of their careers. They will rot unstudied, though. In the unlikely event the runaway net were retrieved by a drift-netting vessel, the strange fish would most probably be discarded. Indeed, given current trends, many species of large, rare fish will probably become extinct before anyone can study them.

The net would have been torn from an industrial fishing vessel. The ship might have flown a flag, but the nationalities of its crew would not much matter. Many countries do not care what their vessels do, especially in international waters. Some countries even make it a business to lend their flags to fleet owners who want to break their own countries' laws. Thus, safety is low, salaries are low, and the motley crews, many without training and experience in fishing, couldn't care less about the long-term state of the resource they are paid to exploit.

The vessel might have been constructed in a subsidized shipyard, to keep jobs in a depressed region of the home country. Or it could have been imported, second-hand and tax-free—another form of subsidy—from a country with a "buy-back" scheme, which allowed that country to modernize its fleet. Though it may be rusty and battered, its electronics are state-of-the-art. The geo-positioning system enables it to pinpoint its position within a few meters, thus enabling it to return to the exact same spot and repeatedly trawl the same productive reef until there is nothing worthwhile left to catch. It can fish in bad weather, even in icy winters. It can travel great distances, for months on end, and thanks to its blast freezers, return with its catch in prime condition. What it cannot do is avoid the results of its own success: a rapidly dwindling supply of fish in the sea. In the last few years, too, its fuel efficiency—the amount of fish caught per unit of fuel burned in its huge diesel engines—has steadily dropped, making it more and more costly to catch fewer and fewer fish.

"Fished out" local waters and upwardly spiraling fuel costs are what drove the Andrea Gail farther and farther from Gloucester. It needed a big catch to justify the rising cost of steaming to evermore distant fishing grounds and staying at sea for weeks on end. When carbon or energy-based taxes are finally put in place to combat global warming, vessels like this will cease to be economically viable. The point may be moot, though, as the fisheries on which it depends may long since have collapsed—and vessels like the Andrea Gail will simply be retired, not swallowed by the North Atlantic.

* * *

Of course, this is not the kind of picture that interests either Hollywood or the fishing industry. Typically, the misleading image they paint is one of relative abundance in the North Atlantic, one in which fishers may have been through hard times but are seeing local stocks rebound and global catches increase; one in which large-scale commercial fishing remains a viable, sustainable enterprise. But in fact, the scientific evidence does not support that rosy picture.

This book presents the best evidence we have on the status of North Atlantic ecosystems. It shows with overwhelming clarity that they are much more like the underwater view presented above than any vision of rebounding health. This evidence, both old and well known, and new, is the result of a synthesis on a scale never before attempted.

In physics, when scientists reach the limit of their instruments' resolution and encounter phenomena that are blurred, they build a bigger machine, a bigger microscope, telescope, or particle detector, capable of clearly detecting the phenomenon under discussion. This decides the issue of its existence, and if it is found to exist, its features.

The health assessment of the North Atlantic fishery is like that "blurry" phenomenon. A debate has been going on for decades about the relative impact of fisheries on marine ecosystems, compared with the impact of the "environment" or "pollution." What we have done through the project documented in this book is, in a sense, to build a bigger machine, one covering not a single bay or gulf, but an entire ocean basin. It covers not one fish species of interest, but all species in the North Atlantic, and especially all species of "table fish," those we like to eat. It encompasses not three or five years of hard-to-interpret, fluctuating abundances of this or that species, but rather 50, even 100 years, thus forcing us to confront long-term impact trends of the fisheries.

This approach is new, though in retrospect, it appears the obvious one to take. On the other hand, the data we incorporated in our syntheses are not new. We did not go out at sea to deploy sophisticated new sensors or survey the entire North Atlantic. Rather, we paid respect to the work of our colleagues in government, industry, and academia, who in the last hundred years have contributed a huge amount of field data, documented in thousands of reports, books, articles, scientific papers, and electronic databases. We developed methods to validate and synthesize their work and use it to document

long-term trends in the status of North Atlantic ecosystems. However, to ensure that our book remains readable, we have included
only our key results in the main text. Readers interested in the background of these results should consult the endnotes of this book and
the literature cited therein.

For two centuries, scientists have been dissecting nature into
ever-smaller pieces. This worked well at first, but as we show, it is
now necessary to re-assemble the pieces back into a larger picture.
Seeing the whole of our watery planet on a single photograph gave
us a new perspective—a new mental map[1]—of its finite nature, its
value to us[2] and especially, its fragility. This new view has enabled us
to visualize environmental issues in the larger dimensions required
to understand more about our planet. It has helped us to understand
how we punched an ozone hole into our planet's atmosphere, and
how we are now changing its climate.

The research described in this book will eventually reach the
same global scope as the work that led to our understanding of
human impacts on the atmosphere. Thus, we are looking far beyond
a summary statement of how healthy or endangered these
ecosystems are—far beyond a one-off assessment. Indeed, our aim
was to create a dynamic methodology that will enable future assessment and monitoring of the entire North Atlantic, and which can
later be used for the whole of the world's ocean.

Nevertheless, the view of the ocean emerging from this synthe-
sized research is not a privileged one, not the only "true" view. Fishers know a great deal about the sea from personal experience, and
their personal knowledge usually exceeds that of marine biologists or
fisheries scientists. Sometimes, though, personal knowledge, which
is inherently local, can be an impediment, preventing one from
seeing the larger picture. Thus, personal knowledge of where to find
cod on the Southern Grand Bank does not necessarily lead to a qualified opinion on, say, the present state of the ecosystem off New
England compared to its state a hundred years ago. Moreover, the
burden of repaying a bank loan invested in the purchase of a better
boat will tend to make it difficult to accept negative conclusions

from scientists whose recommendations, if implemented, would lead to one's boat being retired from fishing.

The trends presented in this book, on the other hand, should be clear enough to the nonfishing public that is interested in the ocean, who might be wondering why their tax money should be used—as it is now—to subsidize a form of fishing which leads to the destruction of the biodiversity of the sea, and ultimately, to the destruction of the very resource base upon which fisheries and fishing communities depend.

Our ingenuity, though, is a real problem. Over the millennia, we have developed the skills required to hunt down even the largest and most powerful animals. This began on land. Except in Africa, where humans evolved, and where the large mammals learned early to fear and avoid us, the largest were hunted into extinction within a few hundred years of the appearance of humans. Thus disappeared the giant kangaroos and wombats of Australia 50,000 years ago, and the elephant-like mastodons and the sloth of North America about 13,000 years ago. The story was similar for large island birds such as the moas of New Zealand, exterminated within a few generations of the arrival of the Polynesians later known as Maoris.

In the North Atlantic, deliberate hunting of whales—as opposed to the occasional consumption of stranded individuals—was probably initiated in Scandinavia a thousand years ago, and then further developed by the Basques, who could practice on the whale populations in the Bay of Biscay, now devoid of large cetaceans. Whaling grew to engulf the entire North Atlantic, and the Atlantic gray whales, probably similar in habits to their cousins along the coast of the northeastern Pacific, went extinct in the process, while the right whales—those that conveniently floated on the surface after being harpooned to death—retreated to the high Arctic, their population now so small that survival of their species is unlikely.

The massacres went on until the mid-20th century, finally spanning the world ocean all the way to Antarctica, where the last major whale populations were fed into the well-oiled machinery of the whaling industry. The revulsion this engendered now protects many whale populations from further exploitation, though some countries continue, albeit under the disapproving gaze of the international community. Some of the whale populations have recovered, but many have not.

Unfortunately, this protection does not extent to fishes, still nearly exclusively perceived as "stock" available for harvest, rather than as the wildlife they are. Hence, the case for protecting fish must be made on economic grounds, as a measure required to avoid the waste of a profitable resource. As it turns out, the case is easy to make: fish biomass in the North Atlantic, i.e., the amount of fish available for use, like savings in a bank, has been severely depleted during the 20th century, particularly following the post–WWII expansion of industrial fisheries. Hence catches, corresponding to the interest obtained from the savings, have continually dwindled. We are now depleting the principal.

In the process, the productive capacity of the ocean has been severely harmed, and there is talk of the damage being irreversible. It is not—except for the extinct species, which nothing can bring back. However, it is becoming more and more difficult to conceive of the abundance that was lost, given the present scarcity, and for many, restoring that past abundance has become a pipe dream. And indeed, it will be more costly to restore depleted ecosystems than it would have been to enforce timely limitations on the growth of the fishing industry. Now, we will have to roll it back.

We ought as well to be aware of how our eating habits impact the sea. Sitting back as we read or watch the world news, or drive a car, or even as we walk down the aisle of a supermarket, it is hard to imagine that as individuals we are an integral part of the food web of marine ecosystems. The division of labor that came from the industrial revolution removed more and more people from direct participation in agriculture or fishing. Mechanization made the proportion of food producers, farmers, and fishers in the population even smaller.

Visiting the countryside in an industrialized country does not get us much closer to our ecosystem roots either. Generally, the view is of huge fields, empty of people, with perhaps a large harvester in the distance; or endless rows of trees placed at such regular intervals that they make us dizzy as we drive by. On the coast, we are unlikely to see a fishing boat or, if we do, it will usually be a large vessel in the distance. The bustle of the food market brings us into closer contact with food production—we may even smell the fish—but the modern market with its clever displays is more a celebration, a seductive exercise in marketing, than it is a demonstration of our utter dependence on food. Here, it proclaims, is the bounty of the land and sea; never mind that some of the beautifully displayed marine products today are actually in danger of disappearing tomorrow.

Rural areas in developing countries are more revealing of our place in the ecosystem. The fields are smaller and the distant machinery is replaced by groups of workers nearby; their implements may be buffaloes or donkeys. On the coast, there are always fishers in view and their boats are generally small, many without engines. The markets are often grim reminders of the increasing difficulty of wresting food from the land or sea, and items are bundled for sale in small quantities. There is a greater sense of the reality of our dependence, almost to the point of urgency, on food and its production.[3]

Though relatively few places are fenced off from human impacts on land, in the sea the distinction between human and wild disappears altogether. We cannot partition and cut up the ocean; we cannot divide the ocean between "ours" and "theirs." It is too far from visible land borders to be easily subdivided, much less defended from intrusion. It has to be shared. By extracting fish from an oceanic ecosystem, we become part of it.

It is critical that we appreciate all that this means. We humans, as the "new predator on the block," can take virtually what we like from the sea, and whatever we remove is taken away from other predators. Unlike us, though, the other predators in the system are unable to turn to other parts of their own food web and certainly cannot turn to rice or wheat or potatoes when marine prey are all gone. On our part, over the course of several centuries we have removed nearly all the large whales (with no rebound in most of their North Atlantic populations) and are presently eating into the populations of the other top predators, the sharks and tunas, and of fishes lower down the food web such as cod and other groundfish, all much reduced from their initial abundance.

Neither are we taking slices out of food webs as if they were apples or cheese, leaving intact the composition and flavor of the remainder. It is more like taking large chunks out of the ingredients of an established recipe: not only is there less to go around; but also the resulting composition and flavor become very different from the original. In terms of ecosystems, this means that different species will thrive, others diminish or perish, and the relationships among them all will change. It also means that for the top predator in the system—us—there will be less fish of the kind we like to eat.

The work and concepts documented herein describe a fascinating adventure. To assess the present status of the North Atlantic's ecosystems, we had to understand their conditions before fisheries began to change them. To improve the present status of the ecosystems, we must aim to return as far as possible to a state that resembles those former conditions. This is actually attainable. We can identify sets of policies and institutions that have proven they are up to the task. The problem is mustering the political will to implement them. But first we must recognize the scope of the problem.

In a world where the past is regarded as firmly "dead and buried," where the terrestrial landscape is everywhere being cut-up and paved over as a consequence of human population pressure and increasing demand from each of us, we have the opportunity to reverse the clearly declining nature of the planet's oceanic ecosystems. The mapping approach that forms the basis of much of this book enabled us to learn a great deal about the history of the ecosystems and the consequence of present trends and alternative management measures. This also enabled us to prepare a "report card" that grades the various aspects of the North Atlantic investigated in our study compared to the relative abundance 50 and 100 years ago. The question we aim to answer, for the North Atlantic and subsequently the rest of the world's oceans, is: How close can we come to attaining a perfect score: A perfect ocean?

IN A PERFECT OCEAN

CHAPTER 1

A Brief History of the North Atlantic and Its Resources

*Unmarked and trackless though it may seem to us, the sur-
face of the ocean is divided into definite zones, and the pattern of the surface water controls the distribution of its life.*
—Rachel Carson, *The Sea Around Us*, 1961.

*. . . the number of the cod seems to equal that of the grains of
sand. . . . These are true mines, which are more valuable, and
require much less expense than those of Peru and Mexico.*
—Charlevoix, 1720s.[1]

The Atlantic Ocean as we know it is only about 20 million years old,
geologically quite young, and its breadth is still increasing by a few
centimeters each year.[2] The Ocean developed from the splitting up
of Pangaea, the only land mass, or continent, in pre-Jurassic times.
At that time a single giant ocean, Panthalassa, surrounded the land.
Pangaea began to break into northern and southern segments about
200 million years ago, and a fissure that is now the mid-ocean ridge
began to divide the American continent from what became Europe
and Africa, creating the Atlantic Ocean between the two landmasses
(Figure 1).

Along with a slow but continuing expansion, there are environmental processes operating in the North Atlantic Ocean at different

1

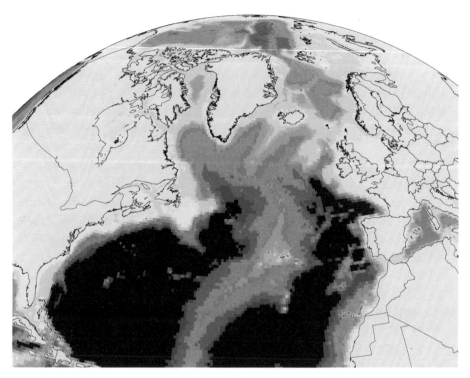

FIGURE 1. THE NORTH ATLANTIC OCEAN BASIN.
Bathymetric map of the North Atlantic as defined in this book. This shows shallow areas (down to 200 m or 600 ft) around the landmasses, where about 90% of fish catches originate, and the deeper area, supporting tuna, billfishes and other oceanic fishes. Map by Reg Watson, based on data from U.S. National Geographical Data Center's Global Relief CD-ROM.

time scales.[3] Long-term processes such as climate change include the ice ages, the last of which occurred roughly 15,000 years ago, and caused northern North America and northern Europe to be covered under one kilometer (one-half mile!) of ice. This greatly reduced sea levels, to the extent that shelf areas such as the Georges Bank off New England and the North Sea were exposed, covered by extensive forests. Medium-term processes last for periods from a few weeks to a decade, and include the North Atlantic Oscillation,[4] which affects the weather in North America and Europe, and can be presumed to affect, as well, the marine "weather" to which fish are exposed, and which, jointly with parental population size, determines their reproductive success.

Finally, short-term processes occur on daily, seasonal, or annual cycles, including tides, which mix nutrient-depleted surface waters with subsurface water, and seasonal upwelling, where wind-driven current parallel to coastlines forces deep water toward the surface. These processes are important, as it is only through mixing of the nutrient-depleted surface waters with nutrient-rich subsurface or deeper water that the nutrients (nitrogen, phosphorus, etc.), which are required for the growth of the algae at the base of marine food webs, are renewed.

The major currents which, from a satellite's-eye view, describe huge arcs, swirls, and eddies across thousands of kilometers, define the oceanic ecosystems of the North Atlantic, while its bottom features define the ecosystems on shelves down to depths of 200 m (600 ft). The major surface current systems in the North Atlantic travel broadly clockwise; those in the southern hemisphere move counterclockwise. This results in very different climates on the east and west sides of the Ocean—the relatively warm water of the Gulf Stream flows the whole year north past the British Isles, while the sea off Labrador at the same latitude is frozen for half the year.[5]

The world's oceans are divided into 4 biomes: the Polar biome, containing polar and subpolar oceans, which make up only about 6% of the total; the Westerlies biome, containing the temperate and subtropical areas of the oceans, about 54%; the Tradewinds biome, corresponding roughly to tropical sea areas, 33%; and the Coastal Boundary biome, comprising all the shelf waters adjacent to land masses, which constitute the remaining 7% of the total ocean area. Globally, these biomes are further subdivided into 57 dynamic biogeochemical provinces—"dynamic" because their borders vary seasonally, and "biogeochemical" because the living organisms ("bio-") therein respond to local ("geo-") processes that determine delivery of nutrients ("chemicals") to the sunlit surface waters, and hence determine the intensity and duration of primary production. The North Atlantic contains 18 of these 57 provinces.

Most of the North Atlantic's provinces are in the open ocean, where the surface waters are infrequently enriched with nutrients from deeper waters. This leads to a low production by planktonic algae, and a generally impoverished environment, similar to terrestrial deserts, inhabited only by large fish such as tuna, which are capable of quickly crossing their large unproductive expanses to find scattered patches of high production "oases."

In contrast, the provinces of the coastal biome, an area of strong water mixing, support high levels of primary production, and it is from the ecosystems embedded in these provinces, e.g., Georges Bank or the North Sea, that most North Atlantic fisheries catches are, or were, taken. These coastal areas, which extend out to the seaward boundary of continental shelves and the outer margins of ocean current systems, can be divided up into "large marine ecosystems" (LMEs), regions of ocean space with distinct bathymetry (the oceanographic equivalent of topography) and productivity patterns. Fourteen LMEs cover the coastline boundaries of the North Atlantic. Fortunately, there is great congruence between the LMEs and coastal biogeochemical provinces, enabling data from both sources to be jointly mapped in a rigorous manner using geographical information systems (Figure 2).[6]

Within each of these larger ecological units, or ecosystems, the numbers and types of fish, the fish "communities," are unique or at least distinctly different from those in other ecosystems. (Note that the term "fish" is used in this book to mean organisms fished: all the animals caught by humans, including fish per se, i.e., "fin fish," shellfish such as crabs and shrimp, mollusks such as oysters, clams, conches, and squid, and other invertebrates like sea urchins. Seaweeds and other algae, although sometimes included among "fish," are not covered here.)

The communities of each ecosystem are interrelated through unique and complex food webs. The animals prey on each other; some marine mammals and sharks prey on large fish such as cod and mackerels; the large fish prey on smaller fish such as herrings and anchovies;[7] and these small fish prey generally on animals among the plankton—composed of tiny less-mobile animals and plants (algae). Other small fish and the animal plankton (zooplankton) eat the plant plankton (phytoplankton), which constitute the bottom of the web. These microscopic phytoplankton use the energy in sunlight in a process called primary production to produce new matter in the form of themselves, as plants do on land. As on land, the sun provides the energy that drives the system.[8] Humans earlier impacted marine food webs at the level of marine mammals and sharks, i.e., as top predators, but the versatility that made our species so successful on land has proven equally effective in the oceans, and we are now attacking all parts of marine food webs, right down to the plankton.[9]

Thus, when fishers extract fish from an area of the sea, they affect the properties—the balance or dynamics, and even the physical structure—of the ecosystem in that area to some extent, in some cases much like the changes in terrestrial ecosystems when clearing a forest for farming. Therefore, fish captured for human use needs to be seen in its ecosystem perspective if we are to assess the real impact of removing the fish. Fisheries science has so far been based mainly on the interactions between fish and fishers, without much attention to the fact that humans (meaning here humanity as a whole, for the fishers are acting on behalf of all fish-eaters worldwide) represent but one predator, added to those naturally occurring in each ecosystem. One of the results, as we will show, is that ecosystem impacts of fishing were not noticed until they had already had their impact on the fisheries.

Past Abundance in the North Atlantic

The earliest accounts that we have of fish populations in the North Atlantic make it clear that both sides of the ocean once contained an abundance of fish—not to speak of marine mammals, seabirds, and turtles—that we may find now almost unimaginable. It is uncomfortable in a way to absorb the fact that the past oceans were so different. Perhaps for this reason, many have doubted the veracity of the historical record. But in fact, the hard paleoecological evidence corroborates these accounts, and careful analysis paints a fairly unambiguous portrait. Past oceans were so well populated with fish and other marine life that the tales of old fishers about their "good old days"—

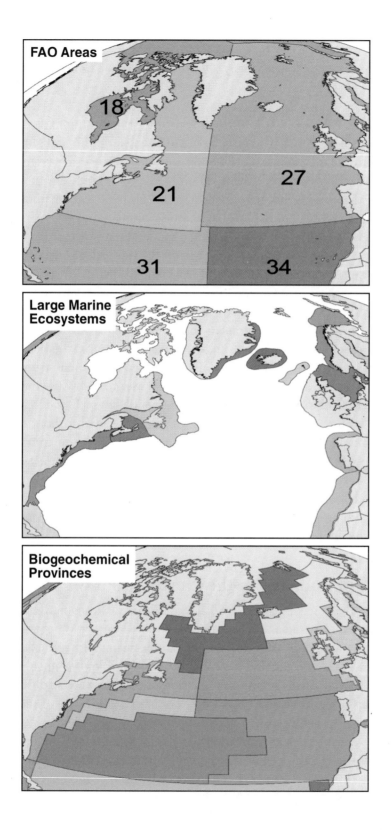

FAO Areas

18

21

27

31

34

Large Marine Ecosystems

Biogeochemical Provinces

FIGURE 2. ZOOMING IN ON ECOSYSTEMS.
Identifying divisions of the North Atlantic suitable for ecosystem-based management will involve moving from the large areas used for statistical purposes to smaller units, e.g., Large Marine Ecosystems, or "Provinces."

The North Atlantic as covered by the *Sea Around Us* Project is made up mainly of FAO area 21 (Northwest Atlantic) and 27 (Northeast Atlantic), but also includes the northern edges of areas 31 (Western Central Atlantic) and 34 (Eastern Central Atlantic), and part of Area 18 (Arctic Sea). Overall, these areas are too large to allow for ecological considerations (see top map). Large Marine Ecosystems (LME) are smaller, and do allow for comparisons among fisheries, but do not cover the central, deeper parts of the Atlantic (and of other oceans; see central map). Biogeochemical Provinces, which largely overlap with LME in the coastal realm, also provide a structure differentiating parts of the open ocean (see bottom map), and thus may provide the framework required for putting fisheries, basin-wide, into an ecosystem context. Maps courtesy of R. Watson based on Pauly et al. (2000), with updates from http://www.edc.uri.edu/lme/default.htm.

that we don't believe as a matter of principle—were more than likely true. Even more discomforting, their forebears would have been saying the same thing, and they would have been correct, too.[10]

On the eastern side of the North Atlantic, archeological evidence going back several thousand years shows that prehistoric hunters in other areas caught a wider variety of larger fish than now. In parts of northern Asia (now the Russian Federation), the salmon and sturgeon harvested were bigger, and there were 40% more species than found there now. Likewise, remains in ancient middens in the Mediterranean contain more kinds of fish and larger individuals than occur in that sea now.[11]

Since at least medieval times, fish were vital for domestic consumption and trade in many countries. It was said a vast army of herrings poured down from the north each year and created the wealth and balance of power among the major European nations. Cod was the staple food on everyone's table, rich or poor, and its relatively low price suggests that it was abundant, hence easy to catch. Yet, in spite of the abundance of fish in their coastal waters, particularly giant cod, halibut, and turbot in the North Sea, the first Europeans to reach the shores of the northwestern Atlantic were amazed by what they saw.

By all accounts, the western side of the North Atlantic was teeming with marine life. Our knowledge of those marine ecosystems comes from "snapshots" from early visitors from the eastern side of the ocean, which make present day reminiscences about the good old days of fishing pale in comparison. Those early navigators, fishers and biographers report that one didn't need a hook, just a basket, to catch big cod, while the ships would bump into whales lazing on the surface, so plentiful were they. The pilgrims on the Mayflower were surrounded by whales when they arrived at Cape Cod. Chesapeake Bay was alive with whales, and settlers discovered it was only one of many bays along the northeast American continent populated by large numbers of the creatures. They were regularly reported to have hindered the movement of vessels. It certainly stretches the imagination, looking at the empty ocean now, to visualize the frustration of these early travelers caught in the whale traffic of the sixteenth and seventeenth centuries. But there was adequate proof of these reports in the massive influx from Europe of fishing and whaling vessels by the hundreds whose crews, with government support, for decades fought relentless if unofficial wars to gain control of the new fishing grounds.

Cod was but one of the many species that made up the wondrous bounty of the northwestern Atlantic. Author Farley Mowat's *Sea of Slaughter*[12] quotes many on-the-spot "snapshots" of the abundance of different kinds of marine life there by visitors in the 1500s and 1600s. Among them were turtles in "inestimable numbers," "an infinite number of them all over the sea"; "harbours writhing with the silver-sided splashing of the stripers" (striped bass); huge salmon in "prodigious quantities"; sturgeons "in great plenty" and "so numerous that it is hazardous for Canoes"; "the greatest multitude of lobsters ever heard of" in the Gulf of St. Lawrence; and a great abundance of oysters and mussels.

Taken in total and even allowing for some enthusiastic exaggeration, the northwestern Atlantic must have been almost like another planet, so alien was it from the previous experience of all these travelers. And it should be noted that in those days, the opposite side of the Atlantic, whence they came, was by no means empty of fish.

Historical reports from the seventeenth century include accounts of divers using diving bells to explore the ocean floor. What might they have seen?

A walk down from the shore into a bay such as the Chesapeake, trailing hoses that supply air pumped by bellows, would have revealed an abundant and somewhat fantastic seascape. A few meters down, fully underwater, there are odd-shaped flat rocks, up to a foot long, lying one on top of the other in vast piles. These are oysters, their shells barely open, gently filtering plankton from the water. In more sheltered places, great clumps of mussels do the same thing. There are so many of these filter-feeders, not to mention dozens of clams buried under every square foot in some sandy or muddy shallows with siphons reaching up to strain the water, that they can keep the bay's waters clear even during the spring blooms of plant plankton. In fact, it is estimated that prior to European settlement, the bay's oysters filtered the entire volume of Chesapeake Bay waters once every three days.

Farther North, in places like Long Island Sound, lurking under rocks and crawling around the bottom, were humongous lobsters of 16 to 25 pounds each. They could easily be netted by the hundreds here, and each could feed several persons. There are plenty of fish sheltering amongst fronds of brown algae that reach upward from the rocks, while sea urchins graze on the fronds. The algae grow fast in these clear waters, fast enough to keep up with the consumption of the urchins.

Elsewhere, there is a wilderness of small animals on or near the bottom, moving around or clinging to rocky outcrops. There are crabs, shrimps, sponges, snails, starfish, squid, and small fish in profusion. At these depths many of the fish are probably the young of large fish like cod, haddock, hake, and pollock. They are protected from predators—including larger members of their own species—by the rich bottom structure, notably sponges and sea fans, providing hiding places from which they dart to catch drifting food morsels.

Depending on the season, there would be huge groups of sturgeon on their way to a river, each measuring up to an astounding 18 feet, bigger than nearly all the sharks and rivaling many marine mammals. Salmon, ranging up to six feet long, swim in massive, shimmering spawning schools as well. Hordes of striped bass, with some individuals weighing 50 pounds or more, patrol near the shore for schools of menhaden; they, too, will travel upriver to spawn. Huge schools of menhaden, capelin, herring, smelts, or other species swim in close formation, almost as one, for protection, eating plankton and becoming food for the larger species. Sometimes loggerhead turtles swim at the surface by the thousand. Abundant leatherback turtles—six feet long, weighing maybe half a ton—troll for slow-moving jellyfish.

Farther offshore the big fish cruise languidly about. Here there are cod, three to six feet long, browsing along the bottom, stopping to crunch a snail or sea urchin or swallow a slow squid or small fish. They are gliding by in every direction, and among them are numbers of tomcod and haddock.

A huge shadow precedes a baleen whale sounding, and another; they fill the whole field of view, underwater and on the surface. The eerie sounds of clicking or groaning are all around as they communicate with each other.

Sudden flashes above are bluefin tuna chasing small fish toward the surface. The tuna are between six and twelve feet long and in a few seconds they are gone. But flashes of other groups pass time and time again, along with the occasional sailfish. What impresses most is the sheer numbers of large fish. No wonder they could be netted or hooked by the thousand, by the million.

And they were.

Development of North Atlantic Fisheries

The view today is quite different. Now, the sturgeons, salmon, and tuna have been all but extinguished; sailfish are rare; only small striped bass can be seen; the schools of smaller fish have been largely decimated. The relatively few lobsters in today's markets are but Lilliputian caricatures of their monstrous ancestors. Turtles are rare visitors. The whales, of course, were nearly wiped out, and the walrus has been extirpated. The birds, nests and eggs have long since been systematically destroyed, island by island, and their largest species, the great auk, or northern penguin, is now extinct.[13] How did this happen?

Bernard Palissy, an early French naturalist (1510–1590), assumed that the fossil fishes known at his time were the remnants of populations "that had been fished too much."[14] While he was wrong about the origins of fossils, his interpretation implies an early awareness of the potential effects of what we now call overfishing. There is good evidence that overfishing, i.e., taking too many fish in a given area in a given time, has been going on in the North Atlantic for several centuries.

The result in earlier times was that fishing fleets developed new gear to access species so far not exploited; or they moved on to find new, more distant, grounds. What ensued is now known as "serial depletions." Successive countries of the North Atlantic rim engaged in this as a matter of acquiring economic power, and even empire building. In fact, the history of fishing and of fish market dominance is intertwined with the political and military history of the North Atlantic rim, with countries succeeding one another in acquiring ascendancy, then having to relinquish it.

Northeast Atlantic fisheries

In the first millennium, the coastline of England was said to teem with fish to the extent that it was hardly necessary to use boats to catch them.[15] Most were taken in various types of traps. However, following the Norman Conquest in the eleventh century, the freedom of the individual to fish any tidal water was withdrawn; nearly all coastal fisheries became controlled by the crown, initially by area and later by type of fish or shellfish.[16] We can speculate that this led to a huge amount of "poaching," as members of communities immediately adjacent to these resources, who had enjoyed near exclusive access to them, found themselves locked out by the armed representatives of a distant king. This issue of "adjacency" is still current, as we shall see below, in the form of tension between artisanal and industrial fisheries.

Herring fisheries were the most important fisheries in Europe from medieval times until the early twentieth century, especially in the North Sea.[17] The autumn herring fishery in the sound at the entrance to the Baltic Sea was the basis of the wealth and importance of the cities that made up the Hanseatic League. Because of this fishery, Copenhagen was established and became the Danish capital.

The Dutch became leaders in the herring fishery in the fourteenth century, and as a result became a major maritime power, continuing to dominate the fishery until the 1700s. Through continuing wars, though, the Dutch lost control of the fishery and the Swedes took over until 1808, when the herring migration pattern changed abruptly and the fishery collapsed. Concurrently, but based on a different population, Norway expanded its herring fisheries, remaining dominant until about 1870. Then, Scotland became the leading herring fishing nation until the early twentieth century, followed by Germany. Finally, England, which had moved faster in adopting new technologies, took over as the prime herring nation.[18]

The traditional fisheries of northern Norway for cod, capelin, and other fish appeared to go into decline after the opening of the country's first whaling station in the 1860s.[19] In the early years of the twentieth century, the cod fishery there collapsed as well, and angry residents, again blaming the whalers, demolished one of the whaling stations during several nights of rioting. Soldiers were called in to restore order. Whaling in the area was banned for ten years from 1904.

Following World War II, technical innovations rapidly improved the efficiency of fishing fleets: acoustic fish finders and radar (products of military research), synthetic netting, mid-water trawlers, purse seine nets, power blocks, improved refrigeration, and huge factory ships that both caught and processed large quantities of fish. These developments, which in turn allowed exploration and exploitation of more distant fishing grounds and fish populations, led to high growth rates in marine capture fisheries, including those of Norway, which ended up collapsing.[20] The spring-spawning herring fishery collapsed at the end of the 1960s, and fishing for them was banned for more than a decade before re-opening under strict control in 1984. Recovery has been slow, or did not occur, as in Iceland, where the fishery for these herring also collapsed.[21]

The Norwegian fishing fleets moved farther offshore following the herring collapse. In the early 1970s, they developed fisheries for North Sea mackerel and herring, for shrimp in the Barents Sea, and for Norway pout and sand eels. About this time the government introduced total allowable catches (TACs) for various fisheries based on advice from the International Council for the Exploration of the Sea (ICES). With the introduction of exclusive economic zones in 1977, Norway shared cod populations with the then USSR. An area closure system was also introduced, in which an area is closed if the proportion of undersized fish exceeds 15 percent, along with the requirement that the mesh size of nets in the cod fishery be increased. Nevertheless, the populations of cod, haddock, and capelin in the Barents Sea collapsed in the mid-1980s, resulting in the closing of the capelin fishery for five years.

Iceland's cod catch fell by about half in 1994. The cod fishery there has since been closely regulated: a total allowable annual catch was set at 25 percent of the weight for fish age four and older. Until 2001, it was an example of an apparently well-managed fishery. According to the estimates from fisheries scientists the stock was showing an increasing trend in recent years. However, in June 2001, an industry e-newspaper, Intrafish,[22] revealed that the cod populations instead were at an all-time low. This led to further restrictions, causing massive losses in investments and mortgages on fishing vessels and equipment. Farther south, fish populations in European Union (EU) waters have been severely overfished for some time. EU officials knew as early as 1990 that fishing should be reduced on almost all the populations examined at the time and, in 2000, concluded that of 67 major fish populations in EU waters, two thirds were overfished and almost half were depleted.[23]

North Sea cod populations have recently been singled out for urgent attention. ICES in November 2000 reported[24] that cod populations in the North Sea were being caught beyond safe biological limits. ICES scientists calculated that a total ban on fishing was needed to restore the other populations. However, the fisheries ministers of the countries concerned reacted with only a 40 percent reduction in allowable catch, although for most years since 1987, fishers have failed to catch their cod quotas due to the scarcity of the fish.

Northwest Atlantic fisheries

Prior to the arrival of European fishers, the western side of the North Atlantic was exploited at a very low level by aboriginal populations, which were themselves small, as imposed by their lifestyle as hunters-gatherers. For example, eastern and northern Canada were chiefly occupied by the Inuits,[25] whose total population was about 9,000 in the mid-twentieth century; the southeast coast was the home of even smaller groups, the Beothuk (Newfoundland) and Micmac (New Brunswick, Nova Scotia). The Micmacs in the early 1950s consisted of mixed descendants numbering about 4,000, roughly the same as the original population. Aboriginals of Newfoundland, of which none now remain, were said to have numbered about 500 in the sixteenth century. Thus, their impact on the marine resources of the vast area of Canadian bays, inlets, and coastal seas was quite small, at least when compared with present practices.

Not only were their numbers low, but also they took only what was necessary for the groups to make it through the winter each year.[26] The Inuits of the Arctic moved inland at the end of spring to hunt caribou or fish in the lakes and moved back to the coast at the beginning of winter to seek seals on the sea ice. They also hunted whales, seals, and walrus from boats with harpoons during summer. The blubber yielded not only food but fuel for light and heating.

It is navigator John Cabot who, in 1497, took back to England an enthusiastic report about teeming fish around Newfoundland, thus revealing the origin of the cod that secretive Basque fishers had been landing in Western Europe.[27] This quickly led to the arrival on the Grand Banks off Newfoundland of competing fleets of English,

French and Portuguese fishers.[28] The British eventually held sway in the area now part of eastern Canada, notably Newfoundland, where Atlantic cod[29] was the major groundfish[30] species. Indeed, commercial fishing interests dominated the settlements.[31]

Movement of the fleets farther north to Labrador avoided the impact of local collapses and the shift from handlines to more intensive catching gear, including cod seines, trawls, gillnets, and traps. Expansion into more offshore fishing grounds accompanied the new trap fishery in the 1870s. Meanwhile, a select committee of inquiry heard, in 1862–1863, much evidence of failing cod fisheries, but the Canadian government failed to act on proposed fishery regulations, and instances of conflict followed over the years. The prevailing attitude was that the more intensive gear was more efficient, and that fishers should all turn to such gear and seek new fishing grounds. By the 1870s, constitutional issues and the opening up of Canada's inland resources placed marine fisheries issues on the back burner.[32]

Overall, the Canadian cod fishery catch[33] could be said to have been at a low and relatively stable level from the 1500s to the 1700s (less than 100,000 metric tons per year). Despite the depletion of local cod populations from the mid-1800s, the overall catch increased with the entry of trawling in the late 1800s (catches rose to 200,000 to 300,000 metric tons per year), and it grew rapidly from the mid-1950s with the introduction of otter trawling. Both developments were accompanied by expansion of the fishing grounds.

In the Gulf of Maine area,[34] small-scale, part-time fisheries first sprang up in the 1600s along the New England coast, in which farmers and coastal town dwellers took advantage of seasonally abundant fish and invertebrates to supplement both diet and income. Over time, more full-time fishers joined the small-scale sector and, in the late 1800s, began to diversify and specialize, with lobsters providing a substantial fishery in Maine. Large-scale fisheries targeting groundfish also developed early in this area, using hook-and-line to take cod, haddock, and flatfish. The largest flatfish is halibut. In the early 1800s they were not considered good eating, but a market for them began to develop, and by the 1830s, they were fished vigorously with longlines by fishers from New England and Nova Scotia.

By the 1850s, only offshore populations remained and by the 1890s, demand was being filled almost entirely by halibut from Iceland. Halibut populations in the western Atlantic never recovered. The response to this was development of purse seining for mackerel, a swordfish fishery, and in the 1900s a trawl fishery, which replaced the hook-and-line fishery.

The 1950s and 1960s were highly productive throughout the Northwest Atlantic, and the cod catch, for example, rose to 810,000 metric tons in 1968. Collapse followed, though, for all the major cod populations in eastern Canada in the early 1970s. This led to the introduction of a quota system in the fishery there in 1974. However, subsequent policy decisions driven by flawed scientific advice on the status of the cod populations resulted in the second and disastrous collapse of the six main Canadian cod populations in 1992. The official estimate of fishery-induced mortality was too low. Survey and tagging data later revealed that it had become very high because overcapacity of the fishing fleet resulted in an increasing discard of juvenile cod in favor of larger fish. Perhaps one-third of all juveniles were being caught and discarded,[35] and most died as a result. As a pilot in the Royal Canadian Air Force described it,

> One morning we raised forty or fifty paired Spanish draggers working Green Bank (Grand Banks area). . . . Some of them seemed to have a tail. . . . When we came over them at about 2,000 feet, we saw it was dead fish. There must have been millions of them stretching out astern of each boat that had just hauled its net and was sorting the catch on deck. Undersized fish [haddock] were going over side like confetti.[36]

The number of spawning-sized cod in eastern Canada in 1992 was found to be around 1 percent of that in 1962.[37] Closures of the Canadian cod fisheries meant that by 1995, some 40,000 fishers and shore workers had been displaced, disrupting many coastal communities. The cod populations were thought to need at least a 15-year period without fishing to recover.[38] But fishing does continue, through quotas set to crop the little population increases that do occur. In the Gulf of Maine and Georges Bank, the cod fishery is subject to severe catch limits and rolling closures, meaning that fishers often have to throw back the cod they unavoidably catch and largely kill in the process, to the extent that these fishers routinely find discarded dead cod from other trawlers in their nets.[39]

Other effects of overfishing[40] of cod in both U.S. and Canadian waters in the 1960s and again in the 1980s were to reduce the proportion of large fish, decrease the length of fish at a given age, and decrease the size at which the surviving fish matured, resulting in fewer eggs per female cod as well as a lower fertilization rate.

Many other fish populations in this region are also in a poor state and, even though in 1992 groundfish abundance was at its lowest in 30 years, overall fishing effort in the area was still increasing in the late 1990s.[41]

Effects of Technology

A critical factor in the development of North Atlantic fisheries is the rapid pace of technological change, which has enabled fishers to range more widely, with more precision and efficiency of capture. One of the first innovations in fisheries to cause concern was the trawl. As early as the fourteenth century, English trap fishers became anxious about a new and "subtly contrived instrument" the "wondrychoun." They feared that this early trawl, which came into use around 1370,[42] would destroy the "flowers of the sea," i.e., the living bottom structure that we now know consists of the body of various animals and which provides shelter, and often food for young bottom fish. The crown concluded that the trawl should be used only in deep water but it was not banned. In the 1620s, King Charles I was forced to consider "the great destruction of fish" caused by trawling and the argument continued into the nineteenth century. The trawl always won the day, but controversy on its effects on the seabed and fish that are not targeted continues up to the present.[43] Indeed, ecosystem effects of fishing are now recognized by the scientific community as a factor that must be incorporated into resource management.[44]

Overall, catches in North Atlantic fisheries during the 1500s to the 1700s increased mainly in direct proportion to the numbers of fishers and vessels. The situation changed markedly with the technological innovations of the nineteenth century. Ice began to be used for preservation in the mid-1800s; steam engines were introduced in the 1860s; steel trawl lines in the 1880s; and otter trawling in the 1890s. These inventions revolutionized fishing, enabling the same number of boats, with an even smaller crew, to land far more fish than previous vessels.

As noted above, following World War II, which completed the transition to diesel engines, various innovations further improved the efficiency of fishing fleets: acoustic fish finders and radar, synthetic netting, midwater trawlers, purse seine nets, power blocks, improved refrigeration, and huge factory ships. With these developments, marine capture fisheries grew at an average rate of around 6 percent annually during the 1950s and 1960s. However, growth in global fisheries fell to about 2 percent in the 1970s and 1980s, and became negative in the 1990s despite continuing refinements in gear technology.[45]

The response of the fishing industry to the falling catch rates that this implies was to incorporate even more technology into its operations. Notably, the end of the Cold War made accessible high technologies previously restricted to military applications, for example precise satellite navigation and positioning, high-resolution maps of the sea bottom, and sensors attached to trawls and other gear. Jointly, these allow finding and catching any fish concentration, anywhere, including under the ice and in deep, rugged areas, and undersea canyons. Fishing was previously impossible in these areas, and they had served as natural reserves for exploited species, allowing a limited number of highly fecund, old individuals to produce the young that sustained the remaining fisheries.[46] With these individuals gone, these fisheries, too, can be expected to collapse, a result of the entire North Atlantic having become a single, giant fishing ground (Figure 3).

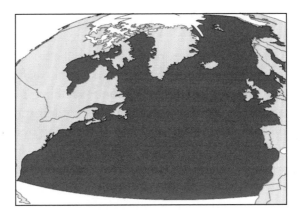

FIGURE 3. DANGER EVERYWHERE!
Map showing, in green, those areas of the North Atlantic where no fishing whatsoever is occurring, while the red areas are those where at least some fishing is permitted.

The point we make here is that, given the resolution of our maps (cells of 1/2 degree latitude and longitude, corresponding to 30 miles x 30 miles at the Equator), only the red color shows, because the few "no-take" marine reserves in the North Atlantic are tiny. Map courtesy of R. Watson, *Sea Around Us* Project.

The Importance of Understanding the Past

The past often holds the key to understanding the present problems of ecosystems and, therefore, perhaps the key to their recovery. For instance, the Chesapeake Bay was for a time the most productive estuary in North America. But today, the only increase in biological productivity in the bay is from plankton, including bacteria. The phenomenon is eutrophication—accelerated growth of algae and plant material as the result of an excess of dissolved nutrients, which depletes dissolved oxygen in the water. In the Chesapeake, the evidence is quite clear in the form of a population explosion of toxic microbes.

This eutrophication is often thought to be the result of farm runoff and other forms of pollution. But recent evidence points to overfishing of oysters that began in the nineteenth century and the physical removal of huge oyster beds as the root causes. As we noted above, millions of these oysters populated the Chesapeake, and they were as large as dinner plates in those days, acting as filters, eating enormous amounts of plankton.[47] Without examining the historical and paleoecological record, we wouldn't know their importance to the health of the bay, and the urgency of restoring oysters to past levels of abundance.

The reference to "giant" cod above is no exaggeration either. On both sides of the Atlantic, they were once magnificent fish. Evidence dating back up to 2,500 years from middens along the northeastern American coast ago shows that cod were regularly caught at lengths up to six feet. This was true in the Gulf of Maine right up until the 1920s, when mechanized trawling began there and both size and numbers of cod fell; the few adult cod remaining now average a little more than a foot. Their decimation in the gulf led to a rise in populations of their prey, including sea urchins, and the sea urchins in their turn destroyed the kelp forests by overgrazing, only to be themselves depleted in recent years by fishers. In the absence of the urchins, the kelp forests, which had survived for at least the previous 5,000 years, have returned.

Such unexpected effects on ecosystems by historical overfishing of individual species are not unique to the North Atlantic. For example, it is probable that diseases recently causing mass mortality of sea grasses in Florida Bay are the result of lack of grazing by green turtles. The turtles' grazing acted to stimulate and clean the beds. Early hunting records indicate a pristine population of at least 40 million adult green turtles; today, they number a few hundred thousand at most. And the sea-grass beds had apparently been present in the area for thousands of years.

West Atlantic coral reefs, stable in their species assemblages for tens of thousands of years, underwent mass mortality in the 1980s, and the dominant corals growing back are now different species. The cause can be traced to loss of a very abundant sea urchin species that itself underwent a mass die-off. The urchin was one of the major consumers of large algae that compete for space with the corals. The other algae eaters, mainly herbivorous fish, as well as sea urchin predators, were fished out beginning in the nineteenth century. When the last remaining controllers of the algae, the urchins, died off, algae smothered the corals.

Overfishing is thus a primary cause of ecosystem disruption, the beginning of chain reactions that reverberate over time—years, decades, or centuries—throughout the system. Without this historical viewpoint, often from archeological evidence, ecologists have come to different conclusions about marine ecosystem changes, often blaming them on changes in water currents, other environmental change, or increased pollution. Where overfishing is not an explicit cause, it is likely to make an ecosystem more vulnerable to other changes wrought by humans. The examples above show how fragile ecosystems are, when the decline of one species can precipitate a major change in the whole system.[48]

The result is that the ocean we encounter today is dramatically different from what it was upon the establishment of the first European settlements in the western Atlantic in the seventeenth century. Our baseline of expectations for what a healthy ocean should look like is not the one described in the beginning of the book, but rather that described in the historical and paleoecological record. It is only against this backdrop that we can truly evaluate the present health of the North Atlantic, the subject to which we now turn.

CHAPTER 2

The Decline of North Atlantic Fisheries

There is, however, a very serious danger, beginning to be recognized, in the case of cod . . . and this is that in time the reckless havoc worked may outrun even these wondrous sources of renewal. Man may by-and-by have to pay dearly for his lack of foresight and commonsense.
—Agnes Giberne, 1910.[1]

Reactions in Iceland [to the disappearance of cod stocks in 2001] have been characterized by shock, astonishment and rage.
—Internet release, June 2001 (www.intrafish.com).

There are clear, and often dramatic, indications of the decline of North Atlantic ecosystems. However, the stark fact remains that until now there has been no clear and compelling scientific assessment of their health. Why not?

With few exceptions, past fisheries research in the North Atlantic has always focused on the fishing industry and on ways to "optimize" catches, or to somehow sustain fisheries that were undermining their own resource base. Generally, attention has been on the large-scale, commercial or industrial fisheries because these have the most economic impact. Catches are usually assessed using models that give a biological maximum sustainable yield of the targeted species or some variant of this yield. The models, mostly developed in the 1950s, assume that a certain, usually constant, fraction of the target species fall prey to other fish; other calculations concern only the relationship between fishing and numbers of target fish. Only one fish species is considered at a time in the process, called a single-species "stock" assessment.[2]

By 1999, it could be said that there was "growing evidence from fisheries around the world that the current scientific methods and associated data used to provide advice for fish-stock management are failing systematically."[3] Our purpose in assessing the North Atlantic on an ecosystem basis was to begin to remedy that deficiency. This chapter presents an assessment of the ecosystems of the North Atlantic as they now stand after centuries of fishing, but particularly since 1950, when the most dramatic advances in technology and capacity have taken place, and when fisheries catch data were first systematically recorded.

Our approach complements, rather than replaces, the single-species methodology briefly described above. Indeed, we relied heavily on biomass estimates from single-species assessments. However, we embedded these in food web models, constructed with the *Ecopath with Ecosim* (EwE) software,[4] representing various ecosystems of the North Atlantic, and verified the occurrence of the prey required to maintain these biomasses in the ecosystems in question.[5] We also used a much wider spatial scale—an entire ocean basin—than is normally used in fisheries research, which usually focuses on specific fishing grounds. As well, our window of study reaches much deeper back in time than is usually done. Our aim was to identify major trends persisting over decades rather than short-term changes. In short, we built a "bigger machine" through which certain phenomena and trends can be established conclusively.

We started by accounting for the depth distribution of the different species upon which the fisheries rely, as well as their distance from the coast. This was necessary because fish tend to occupy different depths and distance from the coast, depending on their life stage.[6] In 1913, F. Heincke formulated a "law," since confirmed by

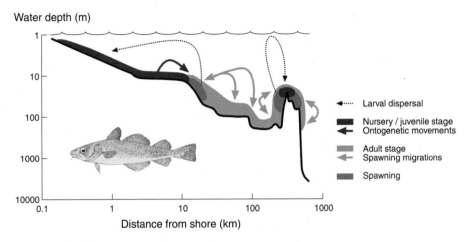

FIGURE 4. WHERE ARE THE FISH?
Populations of cod, and many other commercial bottom-living fish or groundfish, range widely across the coastal waters (continental shelves) of both sides of the North Atlantic. Their movements are basically toward and away from shore to discrete locations at different stages of their life cycle, shown in the example here for Atlantic cod from the Gulf of Maine and Georges Bank. Thus, their life history pattern can be shown in a single "coastal transect," illustrating, in this case, the wide range of depths in which cod are vulnerable to fishing.

This transect emphasizes that cod, like many other groundfish, are susceptible to being caught both inshore by small-scale fisheries and offshore by large-scale operations, and hence the need for coordinating their operations, or at least assessing their joint impacts. Figure modified from Zeller and Pauly (2001).

much research, in which he captured into a neat concept all the mysterious movements of fish from egg to adult across bays, estuaries, and seas. Quite simply, he observed, water depth and/or distance from shore explain most of the life-history distribution patterns of fish. The most common life-history pattern of marine organisms begins with juvenile settlement in shallow water; the juveniles make a gradual transition (or migration) into the deeper, generally offshore, adult area, part of which is a spawning area and from which eggs and young larvae are transported by currents back into inshore waters. These ideas were further developed in the form of "coastal transects," illustrating in intuitive fashion key features of the life cycle of various fish species, and their vulnerability at various stages to fishing

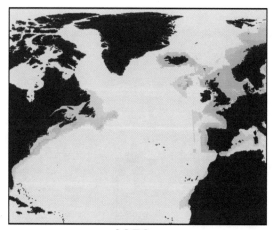

1950s

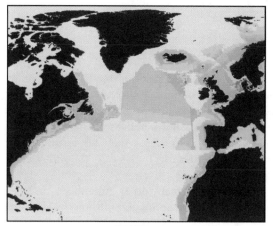

1970s

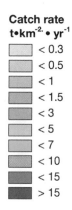

Catch rate
t•km^{-2}• yr^{-1}

- < 0.3
- < 0.5
- < 1
- < 1.5
- < 3
- < 5
- < 7
- < 10
- < 15
- > 15

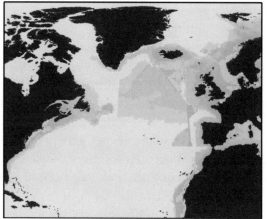

1990s

FIGURE 5. WHERE THE FISH ARE CAUGHT.
Fisheries catches in the North Atlantic, averaged for the decades of the 1950s, 1970s, and 1990s. Areas of highest catches are shown in red, and the lowest in light blue. The catches are mainly on the continental shelves along the edges of the continents. Catches (in metric tons per square kilometer per year) increased to their maximum into the 1970s after more than 500 years of commercial fishing, then decreased in the 1990s.

The maps were constructed using a new method developed by the *Sea Around Us* Project, allowing a new visualization of the process of catching fish. Map modified from Watson et al. (2001a).

apparatus deployed at varying depths and distances from the coast (Figure 4).

Pictures like those in Figure 4 were assembled for the major types of fish, and combined with the North Atlantic depth map (Figure 1) and related information[7] to obtain the distribution of exploited fish populations in the North Atlantic. These distributions, in turn, were used to assign catch records from the various countries (as those countries reported it to the FAO) to the areas where the respective catches must have been taken, in this way generating the first-ever large-scale, high-resolution "catch maps" for this ocean (Figure 5).

These maps enable us to visualize, from an astronaut's-eye view, the fish catches from the North Atlantic Ocean's ecosystems. Notably, we can see that the preference of most fish species for waters not deeper than about 200 m (600 ft) has the effect of concentrating fishable populations, and hence catches, on the relatively narrow, productive shelves surrounding the continents (see Figure 1), leaving the Central North Atlantic as a virtual desert, where only tuna and other widely scattered pelagic fishes can be caught commercially. Simple as it may seem, to create such astronaut's views of the North Atlantic required a tremendous amount of input.[8] The resulting maps represent a completely new way of looking at fisheries.[9]

Having established the particular places in the North Atlantic where the fish live and where they are caught, we then set about trying to establish their biomass in those places. We did this by combining, via a statistical modeling approach, the food web models mentioned above—built on one century's worth of data, from one-third of the North Atlantic—with the spatial catch data illustrated in Figure 5, and maps of the physical features (e.g., depth; see Figure 1) of the over 20,000 half-degree cells covering the North Atlantic. This enabled us to predict fish biomass for each cell and for various years in the twentieth century. The resulting maps, presented below, are at the heart of this book.

Our main conclusion, based on the analyses represented in these maps, related products, and other, independent evidence, is that the North Atlantic Ocean is in a downward spiral with respect to:

- fish catches;
- economic efficiency of the fisheries;
- distribution of fishery benefits to society;
- overall abundance of marine life; and
- the very structure of the ocean's ecosystems themselves.

Of course, the last phenomenon—change in ecosystem structure—is the most important, as it leads to ever greater difficulty in,

---

Box 2.1 Ecosystem

A three-dimensional space within which the interactions between the different residents—animals and plants—are much stronger than their interactions with nonresidents, i.e., the residents of neighboring ecosystems. Ecosystems exist in the sea as on land. In the sea they are not as obvious, even though they are equally dynamic, changing markedly with the seasons. The North Atlantic Ocean can be divided into many distinct ecosystems, e.g., the species-rich, productive shelves, which include coastal areas whose wildlife are impacted by pollution and shoreline development in addition to fishing, or the Central Gyre area of the North Atlantic, a vast, desert-like expanse of low-productivity water, crisscrossed by tunas and other large oceanic fish in search of "oases," the patches of high production where they will feed, but where they also encounter concentrations of fishing gear.

and lesser probability of reversing these trends and of restoring the ecosystems.

In Chapter 1, we introduced the "shifting baseline syndrome,"[10] i.e., the notion, held by each generation of resource users (and natural scientists), that the ecosystems to which they were exposed when young were then more or less pristine, and thus can serve as baseline. Combined with ecosystem degradation within the lifetime of successive generations, this leads to more and more impoverished ecosystems providing ever lower baselines, while the incredibly abundant resources and pristine ecosystems that existed hundreds or thousands of years ago, before fisheries (and agriculture) developed, disappear in the mist of time and really become "in-credible." In recent years, historic and paleoecological research has countered this effect, by describing and quantifying earlier, more pristine ecosystems as rigorously as historic and/or archeological sources allow. We use 1950 as a starting point for most of the time sequences presented in this book, as this is the earliest year with a single standardized set of catch statistics for the entire North Atlantic. However, we have complemented this with a snapshot of North Atlantic fisheries in and around 1900, based on a more disparate set of national statistics.[11]

How Much Fish Is Being Caught?

Reported catches

Combining official catch data from the countries bordering the North Atlantic, we can calculate that presently, total annual fisheries catches in the North Atlantic amount to about 15 million metric tons.[12] Figure 6 shows how this catch has changed over the past 50 years. Catches from the 1950s to the 1970s represent the climax and end of the five-century era of exploration and development of new fisheries and fishing grounds in the North Atlantic.

By the mid-1970s, the annual reported catch reached a peak and has been declining since, with a pronounced dip in the early 1990s due to the collapse of cod fisheries in the Northwest Atlantic, a catastrophic event now threatening Iceland.

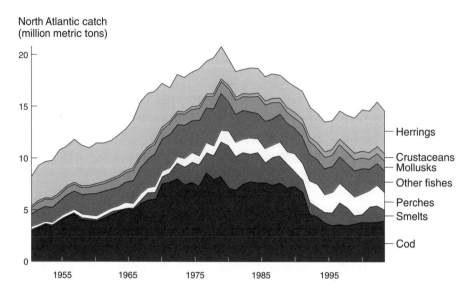

FIGURE 6. THE DECLINING CATCH.
Total officially reported catches of fish in the North Atlantic since 1950. Catches peaked in 1977 after more than 500 years of commercial fishing and are now in steady decline, despite constantly increasing fishing effort.

Based on data from FAO statistical areas 21, 27, 18 (in part), 31 (in part) and 34 (in part). Figure modified from Watson et al. (2001a).

Figure 5, presented previously, and constructed using a new approach developed by the *Sea Around Us* Project,[13] shows the same data, but averaged for the 1950s, 1970s and 1990s. As might be seen, these catches, while gradually expanding offshore, have remained relatively close to landmasses, as one would expect, with more offshore areas being fished in the Northeast than in the Northwest Atlantic.

Discards, illegal, and unreported catches

Apart from the reported catch, there is generally a proportion of discarded fish, often juveniles of targeted or other species, caught as a result of the unselective nature of the equipment used and usually thrown overboard. Some estimates of discards are now available for large fisheries, principally from observer programs in recent years.

The importance of discards can be seen in an assessment in the early 1990s that found the global annual total to be between 18 million and 40 million metric tons.[14] The upper level estimate is half the total marine landings. In other words, up to fully one-third of the catch overall (landings plus discards) is discarded. Figure 7 illustrates the extent of this problem for two representative North Atlantic fisheries.

Other sources of catches besides discards that do not appear in official statistics include illegal catches, fish that are not normally reported ("unmandated"), and catches from sport fisheries.

As an example, a detailed, but preliminary, assessment was made using an adjustment procedure for the cod and herring fisheries on the Scotian Shelf (Atlantic Canada). Here, unreported and illegal catches since the 1960s averaged 30 percent for cod and 175 percent for herring. For the five years to 1999, the values were lower: 15 and 77 percent, respectively, indicating some improvement, but still a continued problem.

Similarly, an assessment of unreported and illegal catches for fisheries along the continental shelf of the northeastern Atlantic, for the period between 1950 and 1998,[15] suggested an average unreported catch of six species (redfish, horse mackerel, haddock, North Sea herring, witch flounder, and European hake) exceeding 20 percent of the total (reported plus unreported). For most of the six, unreported catches were at least 50 percent of the average total catch over each decade. The estimate of average unreported catch of northeast Arctic cod was more modest at 9 percent, but significant considering that the annual reported catch averages half a million metric tons, and there are indications that the unreported catches of cod are underestimated. Some other information obtained during the "detective" work for this assessment can only be used as a guide for future investigation to confirm or refute.[16]

For some fish populations it is becoming clear that their collapse must be traced to management advice based on incomplete data, particularly relating to the total catch. However, the purpose of estimating unreported and illegal catches in the North Atlantic goes beyond seeking to improve fisheries management; accurate data are

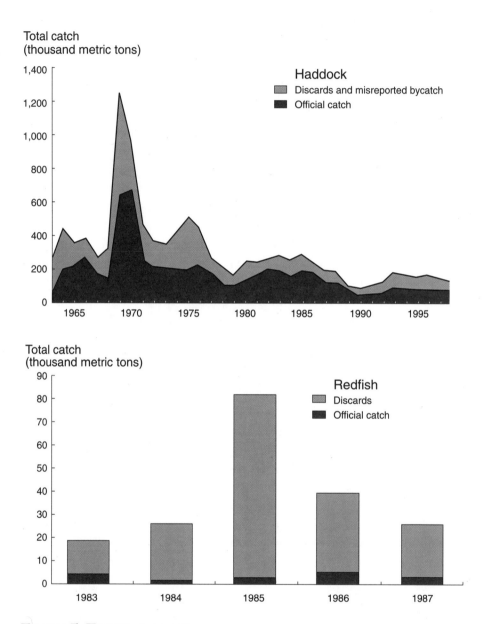

Total catch
(thousand metric tons)

Haddock
■ Discards and misreported bycatch
■ Official catch

Total catch
(thousand metric tons)

Redfish
■ Discards
■ Official catch

FIGURE 7. THE HIDDEN CATCH.
Examples of the discarding of bycatch from North Atlantic fisheries.

Two representative cases are provided: haddock in the North Sea (ICES Areas IIIa & IV), by fisheries from over 15 different countries; and redfish in the Barents Sea (Area I), by Norway and the former USSR. Figure modified from Dingsør (2001).

essential to assess the true impact of fishing on the ocean's ecosystems.

Apart from improving assessment of fish populations and the construction of ecosystem models, information on unreported and illegal catches shows up the deficiencies in the fisheries monitoring and surveillance systems in most countries. Unreported or illegal catches render ineffective what management measures are in place and at the same time undermine confidence in management and management advice. Unreported and illegal catches are certainly a major source of concern, and regulations, if not strictly enforced, do nothing to reduce them. Nevertheless, they can be exposed, as described later.

Fishing Effort and Related Indices

Fishing intensity

Catching fish requires that some effort is expended by fishers. This effort may consist of a few hours standing at a pier in the case of anglers, or of deploying a large vessel for a trip of several weeks. Generally, the more abundant fish are, the less effort needs to be expended to catch them. Hence, measuring fishing effort can help draw inferences on the state of a fish population, especially in combination with the catch itself. Thus, if the catch from a given population is declining, this may be either because the underlying population is depleted, or because less effort is expended. However, when we know that the catch is declining *and* that the effort has increased (or remained constant), then we can infer that the underlying population is declining, and may be in trouble.

Unfortunately, information on fishing effort is even harder to obtain than catch information, and harder still to standardize. How do you compare, for instance, hook-hours and trawling hours, two common measures of fishing effort in longline and trawl fisheries, respectively?[17] In a wilderness of definitions, we show here the overall increase in fishing effort in the North Atlantic using "fishing intensity" as a measure, obtained by dividing catch per unit area (as in the maps of Figure 5) by biomass per unit area, obtained as described below.[18]

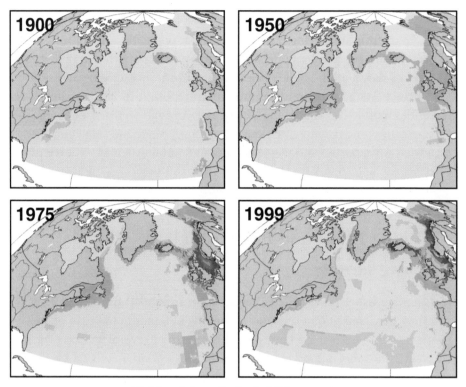

Fishing intensity (catch/biomass)

>1.0	<0.5
<1.0	<0.4
<0.9	<0.3
<0.8	<0.2
<0.7	<0.1
<0.6	<0.05

FIGURE 8. DEATH BY FISHING.

Maps documenting the increase, from 1900 to 1999, of the fraction of large fish killed due to North Atlantic fisheries.

Technically, the map presents "fishing intensity" (i.e., fishing mortality per area) on fish with trophic level 3.75 and higher, as estimated by dividing the mapped catch (as in Figure 7) by the mapped biomass (as in Figure 14). From Christensen et al. (2001, 2003).

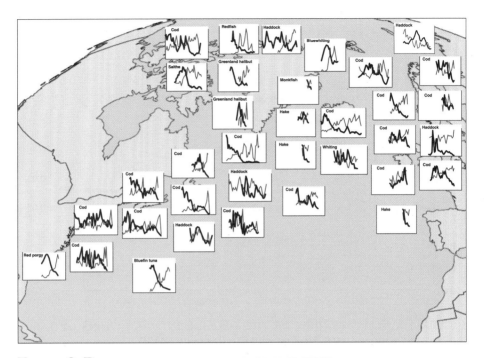

FIGURE 9. DECLINE AND FALL OF A MARINE EMPIRE.
The charts, each representing one fish population, portray the near universal decline, in the last decades, of the abundance of commercial fishes in the North Atlantic (blue lines), and the increase of the fishing mortality to which they are subjected (red lines).

The time scale for each chart is 1950 to 2000. The status of most populations in the 1990s can be seen to have worsened rapidly from what were already depressed levels. Composite graph from Christensen et al. (2001), based on time series assembled by R. A. Myers et al.; see http://fish.dal.ca/~myers/welcome.html.

The result (Figure 8) is a picture of the relentless increase in overall fishing effort that has occurred since the 1950s in the North Atlantic.

As the maps in Figure 8 were derived rather indirectly (from catch and biomass maps, themselves based on indirect approaches), it may be argued that the fishing intensities they present did not increase as much or over such a wide area as it may seem. Figure 9, one of the few exhibits in this book not based primarily on analyses by the *Sea Around Us* Project, shows, however, that assessments of individual populations of predatory fishes (single-species stock assessments) conducted mainly in government research laboratories throughout

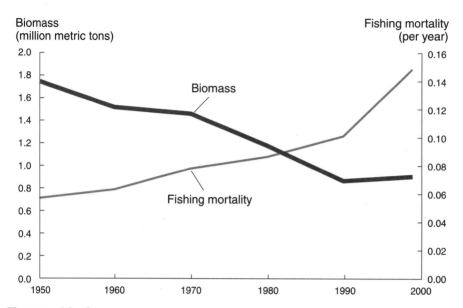

FIGURE 10. SUMMARY VIEW OF THE DECLINE IN POPULATIONS OF LARGE PREDATORY FISH IN THE NORTH ATLANTIC SINCE 1950.
The chart is derived from the detailed catch maps in Figure 5.
    Lines and time period are as in Figure 9. Note the match of these trends with those in that figure, demonstrating that the mapping approach used here for catches and biomass leads to results that are not only comparable with those of single-species "stock assessments," but actually generalize them for the entire North Atlantic basin. Adapted from Christensen et al. (2001, 2003).

the North Atlantic, also found a near universal pattern of biomass de-cline and rapidly increasing fishing mortality, especially in the 1990s.[19]

    Indeed, Figure 10, which presents the biomass and fishing mortal-ity trends derived from our catch and biomass maps, can be seen as representing the "average trend" for the many species in Figure 9.

Economic efficiency

Economic issues, particularly market values, have been and con-tinue to be the main driving force behind most fisheries manage-ment policies. Commercial fisheries in the North Atlantic dwarf subsistence and recreational fishing in terms of catches and overall economic value. Their concerns, which they have been known to communicate quite effectively to decision-makers, have been to land as valuable a catch as possible and to reduce their cost. Yet, paradoxically, information on market values is widely scattered, making it difficult to paint a coherent picture of the value of each fishery and, more importantly, of each ecosystem, as will be required for future ecosystem-based management.

Market value information also makes a solid foundation for economic and social analyses of such issues as where the economic benefits of an ecosystem go: who are the stakeholders and what portion of the total does each obtain? Another important issue is the value trend of a fishery/ecosystem versus the size of the catch—whether or not they are both increasing has a major bearing on future ecosystem management. The annual gross value of the ocean's fisheries can be simply computed as the product of annual catch times the annual average market prices of the different kinds of fish. Thus, if we simply use fish prices for the different years, the result is a graph broadly similar in shape to that for the catch itself.

However, to compute the real value of the catch, we must use the Consumer Price Index (CPI), which takes into account the decreasing value of the dollar over time and the resulting changes in price.[20] This is shown in Figure 11, which also shows price trends for major fish categories from 1950 to the present. As might be seen, fish prices have increased faster than the CPI, i.e., faster than the average of other food products and other commodities. This is especially true for small fishes and invertebrates (shrimps, sea cucumbers, sea urchins, clams, oysters, etc.), demand for which (for fish meal and gourmet products, respectively) has been soaring. This means that, while the catch may have declined significantly, the value of the catch has not declined as far. This itself implies that the consequences of overfishing are not being felt in terms of overall economics: much fishing effort has moved from depleted populations of larger fish to more lucrative small invertebrates.

That larger fish have been replaced by invertebrates in fisheries catches is a phenomenon which, taken at face value, appears to be a

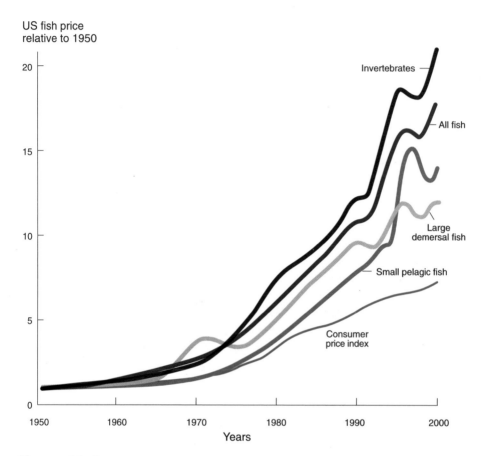

FIGURE 11. INFLATION PLUS.
The rising prices of fish in the North Atlantic Ocean compared with the consumer price index (CPI), which represents the average U.S. inflation rate since 1950. Note that all groups have increased faster than the CPI, implying that fish products, and particularly invertebrates (such as shrimps, lobsters, and scallops) have increased in price faster than the inflation rate, suggesting an increasing demand, which masks reductions in catch values that would be seen otherwise. Based on data in Sumaila (1999).

good thing, as new resources are being used to fill a demand, benefiting both fishers and consumers. Yet it is this very feature—moving from larger predatory fish (as their populations become depleted through excessive fishing) to their prey or even their prey's prey— that is the essence of the North Atlantic Ocean's problems. The process of chasing predatory fish, their prey and then their prey's prey is called "fishing down the food web." The important consequences this has for the ocean are discussed later.

Energy efficiency

Energy costs of fishing[21] usually receive less attention than the direct impact of fishing on fish populations, although it is precisely the availability of abundant, inexpensive energy that allows many fisheries to continue even when populations are declining. Moreover, the use of fossil fuel itself has an increasing, if indirect, impact on marine ecosystems via its contributions to global warming.

Energy consumption data can lead to a number of insights, for example:

- That the energy intensity of particular fisheries and the amount of protein derived per unit of energy input are decreasing, thus providing an independent measure of decline in the abundance of the underlying resource species.
- That the high rate of carbon dioxide emissions of highly mobile fisheries makes them vulnerable to the effects of carbon taxes, likely to be imposed on various industries in the coming decades.

In the days when fishers rowed their boats, the energy intensity of fishing obviously had to be much higher than 1 (that is, if fish was your staple food, unless you extracted more energy from fish than you put into catching them, you ended up starving). Once fisheries began using cheap fossil fuel to replace human labor, this limitation was no longer an issue.

Most of the energy used in fishing is from burning fuel, which typically accounts for 75 to 90 percent of the total, the remainder being made up of energy associated with building and maintaining vessels and the provision of equipment and labor. In general, trawling uses more energy than seining, purse seining or more passive operations such as gillnetting and trapping. Energy use can also vary greatly over time in response to changes in fish abundance and location, expansion of fleets, and changes in vessel size and efficiency.

The energy used in North Atlantic fisheries was assessed in two ways—by soliciting relevant information from fishing companies, and by combining estimates of the generic rates of fuel consumption by fishing vessels in relation to the horsepower of their main engine. The engine data were derived from real-world vessel performance data and fishing effort data.[22] For some gear sectors, such as trawlers and seiners, fuel consumption was found to have little relationship to catch type or rate, and to depend much more on the nature of the vessels and their gear. Comparing fuel consumption and propulsive horsepower for a range of engine sizes established values for these two gear types. The numbers of horsepower-days fished leads straightforwardly, via the fuel consumption of engines, to estimates of the total fuel energy consumed in each fishery.[23]

The fuel energy use of 54 fisheries in five countries—Canada, Germany, Iceland, Norway, and the United States—was assessed here. Total annual production of the fisheries concerned is more than five million metric tons, one-third of total North Atlantic catches. A wide range of fisheries types are covered, thus making the sample representative of North Atlantic fisheries as a whole.

Fuel consumption per horsepower per day at sea ranged between 1.88 and 2.55 liters, the lowest values occurring in standard purse seiners and the highest in vessels using trawls and dredges. The average for nearly 200 vessels represented in the analysis was 2.53 liters, and the total for all fisheries just over one billion liters per year. This translates into over 3 million metric tons of carbon dioxide (CO_2), the major greenhouse gas, released annually into the atmosphere.[24]

Here is a summary of the fuel efficiency of different fisheries, measured as ton of fish caught per ton of fuel burned.[25]

- Small pelagic fisheries: 10 to 50 with an average of 20
- Ground fish fisheries: 4.3 to 5.0
- Two scallop fisheries: average 3.0
- One crab fishery: 3.0
- Eight shrimp fisheries: average 1.2
- One Norwegian lobster fishery: 1.1
- One fishery for large pelagic fish (swordfish and tuna): 0.7

These values change over time. Fuel efficiency values have gradually decreased, with great variability in recent years. For instance, there was a gradual decrease in fuel efficiency in Canadian groundfish fisheries until 1991, when the Atlantic cod populations collapsed, and a subsequent increase to levels higher than in 1986. Icelandic groundfish fisheries have nearly all shown gradual to sharp decreases in efficiency over the 20 years from 1977 to 1997 (Figure 12).

A more precise measure of fuel efficiency is obtained by keeping fuel energy as the input, but by stating the output in terms of the protein energy in the parts of the fish consumed (thus accounting for processing waste, such as bones in finfish, and shells in shellfish).[26]

By that measure, the fisheries for small pelagic fisheries fare best, providing between 30 and 260 units (average 130) of edible protein per 100 units of industrial energy expended. A very different picture is seen in fisheries whose production is directly consumed by humans: in invertebrate fisheries (shrimp, scallops, etc.), the values range between 1 and 13, with an average of only 4 units of protein energy per 100 units of input energy, or 1 in 25, while groundfish fisheries yield from 2 to 25 units of protein energy per 100 units of input energy (average 10). Moreover, the trend, at least in the latter fisheries, is declining as well, as illustrated by data from the Canadian and Icelandic groundfish fisheries, whose output of edible protein energy per unit fuel energy has been declining more or less steadily over the years for which data are available (Figure 13).

The high returns for the fisheries exploiting small schooling fishes appear to be unique. However, their benefits are largely wasted, as these small fishes are mainly used as animal feed, with over 90 percent of their energy contents lost in the process. Fish farming in some locations (e.g., tilapia in Africa or carp in Israel) provides about 10 units of protein energy per 100 input energy units, while in others (catfish in USA, sea bass in Thailand, salmon in Canada), less than half that ratio is achieved. Livestock operations also give low yields: beef, chicken, and lamb examples attain only about 2 units of protein energy per 100 input energy units.[27] Of course, the higher returns from the North Atlantic fisheries shown here cannot be used as an argument to further increase energy input, i.e., fishing capacity, in

Catch in weight / weight of fuel

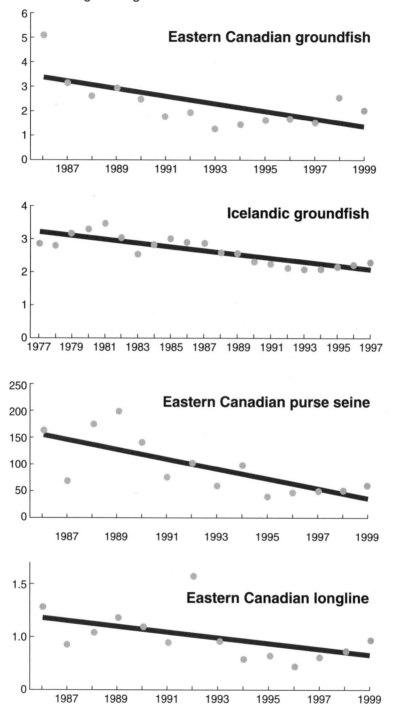

FIGURE 12. MORE FUEL, LESS FISH.
The charts to the left show the trend over time in the amount of fish caught per unit weight of fuel, for a diverse set of North Atlantic fisheries. There is a large difference in "fuel intensity" between fisheries, reflecting the relative value of the targeted species. In all instances, however, a decreasing trend occurs, typical also of other fisheries in the North Atlantic.

The two upper graphs represent trawl fisheries for fish, including cod, used for human consumption. The third graph refers to fuel efficient fisheries for low value, small fishes such as herring and capelin, used mainly for fishmeal and oil, used to feed pigs and farmed salmon. The fourth (bottom) graph pertains to a fishery for high-value tuna and swordfishes. Figures from Tyedmers (2001).

these fisheries. The evidence overwhelmingly implies that the "low-hanging fruits" have all been harvested, and that more energy input is likely to lead to even lower protein energy returns.

Note that, as we fish down the larger fish and the proportions of shrimp, clams, and other invertebrates in the catch increase, the catch contains less protein per unit of landed weight because invertebrates, with their tough shells, have lower proportions of protein than do fish. Overall then, we are getting less protein from the sea over time, expending more energy to obtain the same amount of protein, and incurring higher and higher real costs for that protein.

These trends may be seen as another aspect of the "fishing down the food web" phenomenon, characterized by increasing catches of smaller species, and an increasing proportion of organisms from the bottom of food webs, typically invertebrates increasing over time (see below).

This may be perceived as resulting from a change in public taste (squid were considered bait in many Western countries until recently, but have become a target species prized for human consumption), but should rather be seen, as we shall show below, as an expression of ecosystem change.

Impacts on Biomass and Ecosystems

Before presenting what we believe are the main findings of the *Sea Around Us* Project regarding the ways fisheries impact on the North

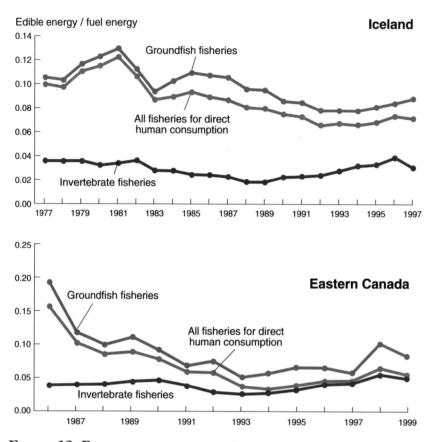

FIGURE 13. RUNNING OUT OF ENERGY.
Changes in the edible energy extracted from caught fish per unit of fuel energy used
in catching groundfish and invertebrates in Iceland and Eastern Canada.

Note declining trends, representative of the entire North Atlantic, and tending
toward ratios of 1:20; the lower edible energy content of invertebrates is largely due
to their shell; smaller fishes also have lower edible energy contents than large fishes,
and hence the gradual convergence between the fish and invertebrate trend lines.
Figure from Tyedmers (2001).

Atlantic, we would like to make an analogy to a weather map. Just as
it is understood what the wisps of white, and a "High" and a "Low"
are, it is necessary to understand two technical terms to interpret
the pictures presented below.

- *Biomass.* This is the amount of fish that is in the sea at any moment in time. It is usually difficult to estimate; you can't count fish as you can trees or even bears in the forest. Biomass is the basis of all food production, and when we catch fish, we reduce their biomass. However, if left alone, biomass will grow, a bit like money in the bank, and we can draw the interest from that capital. Hence the more biomass in the sea, the more interest (or catch) we can obtain, and the better we are buffered against bad years or unforeseen expenses.
- *Trophic level.* This number, ranging from 1 in plants to 4 or 5 in larger predators, expresses the relative position of fish and other animals in the hierarchical food webs that nourish them. Cows have a trophic level of 2: they eat grass (or at least they should). In the ocean, tiny animals—zooplankton—which also have a trophic level of 2, play the role of cows. We do not eat them, but they provide the food for most small fish, which have a trophic level of about 3. Usually, we do not eat small fishes, either. Rather, we generally eat the fish that feed on the small fishes, which have a trophic level of around 4. This means we are at around trophic level 5 when we eat fish. Conversely, we are at trophic level 3 when we eat a beef steak, and at trophic level 2 when we eat a peanut butter sandwich. Note, finally, that the trophic level scale resembles the famously "open-ended" Richter scale for the magnitude of earthquakes, where a level 5 is 10 times stronger than a level 4 quake. Biological production, and catches also change by a factor of about 10 by trophic level.

As the example above illustrates, an animal or person can operate at different trophic levels during the same day, and thus end up with an intermediate trophic level, such as 2.4 or 3.8. Also, we can conceive of temporal changes; for example, bears work at different trophic levels in different seasons: 2 during the berry season, 3 during the period when plant-eating voles are abundant, and well over 4 during salmon runs, thus ending up with an annual mean of, say, 3.3. By switching food in this manner, bears can overcome lean seasons, or even lean years. They can rely on all of the forest's eco-logical content. Fish do the same. Most of the fish we eat feed at several trophic levels and thus can overcome lean periods.

Now, what is the big picture for the North Atlantic in terms of its biomass of fish through time, and in terms of the trophic levels of the fish? We recall from above that in the 1950s, overall catch and catch value in the North Atlantic were low. It was a good thing that the fisheries grew. The western world had been devastated by war, and here was a resource that was underexploited. Catches grew rapidly in the 1960s and 1970s. However, our fisheries management agencies never really succeeded at stopping the growth of a fishery once it started, and the North Atlantic fisheries overshot. They became far too large, and in consequence depleted the biomass severely. Thus, our capital in the bank and our insurance against lean years was drastically reduced, especially at high trophic levels. This led to the process we now call "fishing down marine food webs," which oc-curred throughout the Atlantic—and in all other oceans of the world—in spite of all our regulatory agencies and as a direct result of too much fishing.

The trend, which is steady and broad, leads straight into a situa-tion where the few remaining large fishes and marine mammals must feed at low trophic levels: a few bears are left and all they have to eat are berries. Or put differently: the food webs of the North At-lantic have been drastically shortened, and the whole pyramid of sea life squashed down, forcing us to compete with seals for the few re-maining forage fish.[28]

These are not speculations. The maps that are presented below were not drawn with crayons from a dash of inspiration, but con-structed from diet composition studies of hundreds of thousands of marine organisms, painstakingly standardized and balanced against the food consumption and abundance of thousands of fished and un-fished organisms, derived from thousands of studies conducted by the scientific community during one hundred years of biological and fisheries research around the North Atlantic.

This is the reason why these maps resemble weather charts, as-sembled from the records of thousands of weather stations throughout the world. This is also the reason why the maps in this book resemble those produced by the scientists working on global climate change. What we have here is an integrated picture of the state of an entire ocean. The trend illustrated by these maps will be very hard to reverse because it is not due to one or two causes. But reverse it we must; if not, our seas will soon be full of jellyfish happily feeding on zooplankton. And our fishers, already hard pressed by their own excesses, will have to cease for want of products to land. We now present our case in some detail.

Box 2.2 Biomass

The weight at a given time of all the animals and/or plants of a given population (or age group of a population), or in a given ecosystem. Usually estimated from population numbers estimated by mathematical models, multiplied by the mean weight of representative individuals.

Box 2.3 Trophic Levels

Each type of fish has a certain position in a food web. Plants are at the lowest level (trophic level 1) in a food web, eaten by herbivores (level 2), which are eaten generally by successively larger animals, i.e., at higher trophic levels (e.g., levels 3 to 4) with, in the case of oceans, marine mammals and humans at the highest levels (usually up to 5). Most marine species have mixed diets: a shark feeding only on cod that have a trophic level of 4.0 would itself be at level 5.0, but if the shark feeds equally on a small carnivore of level 3.0, then its trophic level would be 4.5. Most fish have trophic levels between 3.0 and 4.0.

Biomass declines

The total amount of fish in the ocean is declining. We have opted not to give here detailed, species-by-species accounts, representative only of limited areas. Instead, we refer here to the biomass of all the larger fishes on top of marine food webs in the North Atlantic as a whole: the cod, halibut, and other bottom fishes, and tuna, billfishes, and other oceanic fishes, and the sharks that are found in both habitat types.

These fishes, except perhaps for the sharks, are those we typically eat, whereas the other fish of the sea, the herring, sardine, capelin, etc., which feed on zooplankton, are their prey. Hence the fishes we examine here can also be described as predators, or high trophic level fishes, while the others are forage, or low trophic level fishes.

The maps in Figure 14 show how the biomass of predatory fish in the higher trophic levels has been steadily decreasing during the period examined.[29] As might be seen, the red areas, corresponding to high biomasses, (a "High" on a weather chart) declined, then largely disappeared, turning the entire ocean into a "Low," reflective of the bad weather buffeting our fisheries. Obviously, if this trend continues, there must be a point at which more and more fisheries based on high trophic level fish will follow the Canadian and other cod stocks into oblivion.

Now it could be argued that the elaborate map-based methodology used to construct Figure 14 generates spurious trends, distinct from those observed by those studying single species. Figure 10, presented earlier, which summarized much of the single-species work done in government fisheries research laboratories all around the North Atlantic in recent years, showed, at least for the period from 1950–1999, that it is not so: almost all high trophic level species in the North Atlantic suffered sharp biomass declines in the last decades.

Fishing down marine food webs

The impact of fishing on the North Atlantic's ecosystems, illustrated above by missing fish, can also be seen by examining the composition of the fish that are being landed, i.e., via their average trophic level. Because of the close relationship between trophic level and

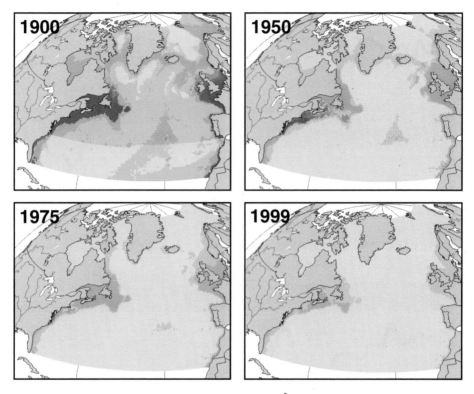

Biomass (t•km^{-2})

>11	< 6
<11	< 5
<10	< 4
< 9	< 3
< 8	< 2
< 7	< 1

FIGURE 14. GOING, GOING . . .

The larger, predatory fish types in the North Atlantic have been decreasing since 1900, especially in areas where they were formerly most abundant (red color).

"Larger fish" here are those of trophic level 3.75 and higher. Their biomass is expressed in metric tons per square kilometer, with red pertaining to biomasses of 10 tonnes and more per square kilometer. Maps from Christensen et al. (2001, 2003).

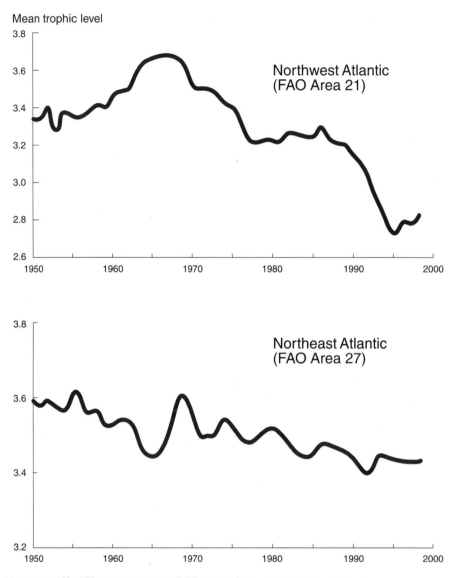

FIGURE 15. "FISHING DOWN" NORTH ATLANTIC FOOD WEBS.
Trophic level trends in the North Atlantic, 1950 to 1998, indicating the rapid (in the West) or gradual (in the East) replacement of large predators in fisheries catches by small fishes and invertebrates.

Various studies have shown that such changes in catch composition indicate changes in relative abundance in the underlying ecosystems, and thus reflect the collapse of large populations (when trophic level declines are rapid) or the serial depletion of a number of smaller populations (when declines are gradual, but continuous over longer periods). Modified from Pauly et al. (1998c)

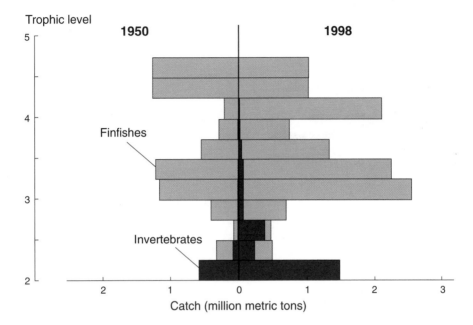

FIGURE 16. DIGGING DEEPER DOWN THE MINE.
The catch composition in the North Atlantic has shifted over the past half century toward fish lower down in the food web, that is, from large finfishes (especially cod) to smaller finfishes and to invertebrates, especially mollusks such as clams.

The figure shows annual catches for 1950 and 1998 in FAO areas 21 and 27, as output by FishBase 2000 (Froese and Pauly 2000); the corresponding trophic level trend for these two FAO areas is given in Figure 15.

size (remember: big fish eat small ones), mean trophic levels reflect changes in both size composition and position in the food chain, and hence ecological roles.

Figure 15 shows that the catches of large (high trophic level) fish in the North Atlantic is declining, relative to that of low trophic level small fish and invertebrates.[30]

Figure 16 provides another view of these changes in trophic levels, in the form of catch pyramids, wherein the composition of North Atlantic fisheries catches in 1950 is compared with that of 1998.

Another way of presenting the "fishing down marine food web" phenomenon is through Figure 17, which summarizes much of what

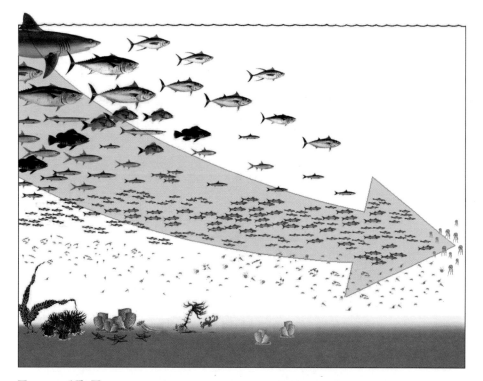

FIGURE 17. FISHING DOWN: WHAT IT ACTUALLY MEANS.
Fishing down marine food webs means that the fisheries (blue arrow), having at first removed the larger fishes at the top of various food chains, must target fishes lower and lower down, and end up targeting very small fishes and plankton, including jellyfish.

In some parts of the world, the fisheries have indeed gone all the way down. In the North Atlantic, most fisheries still operate at the level represented by the center of the figure; however, the trend is down. The invertebrates in the lower left part of the graph gradually disappear because of trawling, which smothers them and turns the sea bottom into vast mudflats (Design: D. Pauly; artist: Ms. Aque Atanacio, FishBase Project, Los Baños, Philippines).

is known about this process. Therein, the fish are arranged as in food webs, with high trophic level fishes high on the vertical axis. Fishing (blue arrow) depletes those fishes, and given time (horizontal axis), their abundance declines, forcing the fisheries to concentrate on fishes with lower trophic levels. This reduces the food base of the predatory fishes, which decline even more, and thus force the fisheries to move further down the food web. This downward spiral then takes us all the way to the lower right corner, where the fisheries, having depleted the larger, then the smaller fishes, will turn to the zooplankton, including jellyfish. In fact, this is already happening:

> "It's tasteless, but it's something chewy that has a crunch to it," said a Georgia (USA) jellyfish processor, talking about his product (Associated Press article, May 2001).

"Too extreme," was a comment from one senior U.S. fisheries scientist on the idea of being reduced to catching and eating jellyfish. The Japanese, more than most, have been eating them for some time, of course. Formerly out-of-work fishers in Georgia, U.S., who are now finding both a lot of jellyfish in their waters—one ex-shrimper said he can catch 50,000 pounds a week—and a ready market in Japan,[31] are beginning to make handsome profits. One can confidently predict that this, and other East Asian markets will grow in the United States also.

Squashed pyramids and shorter food chains

Consequent to the trophic decline and decreasing biomass, the structure of the ecosystems is changing. Figure 18 shows the changes in ecosystem structure over 100 years in two representative areas: the North Sea and the Grand Banks. In both cases, the food web has shrunk—the amount of biomass that is transferred up the food web has declined dramatically, and there is insufficient food at high trophic levels.

The consequence for the predatory fishes that have not been fished out is that they must feed on lower trophic levels, where food remains. This reduces the average number of links in the chain between predatory fish and the plankton. Long food chains are disappearing, leaving mainly shorter ones. The diagram in Figure 19 illustrates this for a typical large predator.

This may appear to be an academic point, interesting only to food web theoreticians, but it has, in fact, important practical consequences. Notably, shorter food webs expose top predators to the

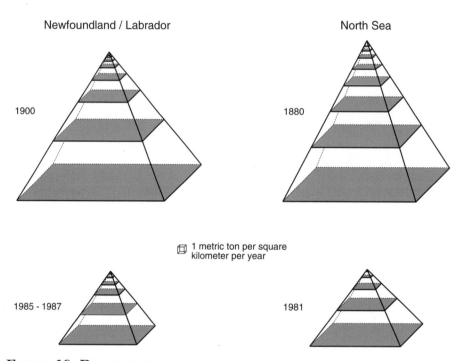

Newfoundland / Labrador

1900

North Sea

1880

⊞ 1 metric ton per square
kilometer per year

1985 - 1987

1981

FIGURE 18. DIMINISHED PYRAMIDS OF LIFE.
Less enduring than their namesakes in Egypt, the ocean's food webs, which can be conveniently represented as pyramids, have been "squashed" by a century of unsustainable fishing. These examples are from the eastern North Atlantic (North Sea, in 1880 and 1981) and western North Atlantic (Newfoundland/Labrador, in 1900 and 1985–1987).

Each layer in the pyramids represents a trophic level and its volume is proportional to the biological production of organisms at that level in the ecosystem. Over time, both the number of trophic levels and their associated flows have decreased dramatically. Figure 19 provides a schematic for the shortening of food webs implied here. Sources: North Sea 1981: Christensen (1995b); North Sea: 1880: Mackinson (2001); Newfoundland/Labrador 1985–1987: Bundy et al. (2000); Newfoundland/Labrador 1900: Pitcher et al. (2002).

strong, environmentally-driven fluctuations exhibited by the plankton organisms at the base of food webs, fluctuations which were previously dampened by food webs with a variety of strong and weak links.[32] Or put differently: fishes were previously consuming fishes which themselves were consuming various kinds of smaller fishes that consumed plankton. Now, however, the fishes that fisheries

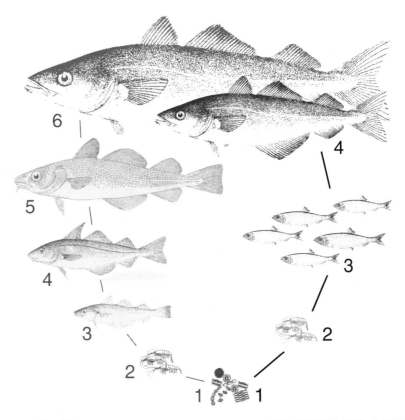

FIGURE 19. WHAT FISHING DOWN THE FOOD WEB MEANS FOR A TYPICAL PREDATOR.

As once-abundant large prey become scarcer, smaller prey must be consumed. This then affects the whole food web, resulting in fewer steps between the extremes of the predators and the plankton.

The numbers represent trophic levels. In the example, in the food chain on the left, large saithe consume cod, which consume whiting that consume haddock that consume kill and/or other zooplankton which feed on the tiny plants comprising the phytoplankton. The shorter food chain on the right consists of smaller saithe consuming herring that feed on zooplankton. In pristine systems, food webs consist of long and short food chains, thus providing stability for the predators. This diversity is absent in the overexploited systems, which are dominated by short food chains. Graph courtesy of Villy Christensen, based on data in FishBase (www.fishbase.org).

target tend to feed directly on a few species of plankton-feeding fishes, and thus are far more exposed than previously to seasonal and between-year changes of plankton abundance. Thus, the fish species targeted by fisheries, earlier very predictable in terms of their biomasses, now fluctuate more widely than before.

This effect, moreover, has been amplified by the scarcity of old specimens in the exploited populations, whose magnitude, therefore, tends to vary with reproductive success, i.e., the entry of young fish ("recruits") into the populations. Combined, these two effects make catches even harder to predict, and fisheries more difficult to manage than they already were.

Competition with marine mammals

Humans are not the only mammals acting as predators in the ocean's ecosystems. Many marine mammals, especially pinnipeds such as seals and sea lions, and the toothed whales, including dolphins, are at or near the top of the food chain. The decline in their numbers is reflected, along with the decline in large fish, in the shrunken pyramids of Figure 18. Even though large-scale commercial whaling ended in 1986, and the extent of sealing has decreased substantially[33] over the past 50 years, many species in the North Atlantic have not rebounded to "push up," as it were, the height of those pyramids. Figure 20 shows the decline in their populations since the 1950s. All baleen whales are currently protected under the CITES international trade agreement, and the majority of them have been listed as endangered or vulnerable in the IUCN Red List.[34] Some, such as the blue, bowhead, and Northern right whales,[35] now number only a few hundred individuals.

Many North Atlantic marine mammal populations remain low following the effects of past fishing pressure, from which many have failed to recover.[36] Part of the reason may well be lack of food due to competition with fisheries.

In the 1950s, the populations of marine mammals in the ocean were eating seven times more fish than was being caught by the fisheries.[37] This ratio dropped to about three in the 1990s as a result

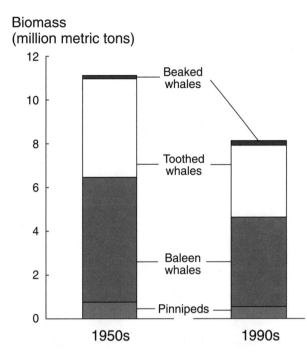

Biomass (million metric tons)

FIGURE 20. REDUCED POPULATIONS OF MARINE MAMMALS.
Marine mammals in the North Atlantic, apart from the relatively rare beaked whales, have declined since the 1950s.

This contrast may be surprising in view of the emphasis that recent increases in a few populations of pinnipeds have received in the media. Overall however, marine mammals are still less abundant than in the 1950s. Adapted from Kaschner et al. (2001).

both of decreasing marine mammal populations and increasing fisheries catches (Figure 21). But when only the types of fish taken by fisheries are considered, it emerges that marine mammals are now eating about as much as we catch for our own consumption and processing into animal feeds. Considering the documented downward trends in marine mammal biomass and the increase in fisheries over the past fifty years demonstrated in this study, it does seem to indicate that some fishers and fisheries managers are using marine mammals as scapegoats for their own mistakes.

Figure 22 shows where in the North Atlantic the different groups of marine mammals obtain their food.

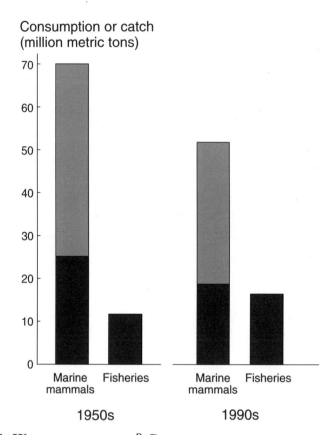

Consumption or catch
(million metric tons)

FIGURE 21. WHO CATCHES WHAT? COMPARING THE FISH EATEN BY MARINE MAMMALS AND HUMANS.
Overall, marine mammals in the North Atlantic now consume less food than they did in the 1950s, and less of the kinds of fish that we actually catch.

Also note that in both time periods, the majority of marine mammal food consists of items (turquoise) not targeted by fisheries. Adapted from Kaschner et al. (2001).

As the distribution of marine mammals in the North Atlantic is much broader than human fishing grounds, much of their food comes from areas other than where the fisheries operate. Thus the spatial overlap between fisheries and marine mammals is quite small if we consider the North Atlantic as a whole.

The areas and extent of "diet" overlap between fisheries and certain groups of marine mammals, especially the pinnipeds and baleen whales shown in Figure 23, does indicate, however, a few "hotspots"

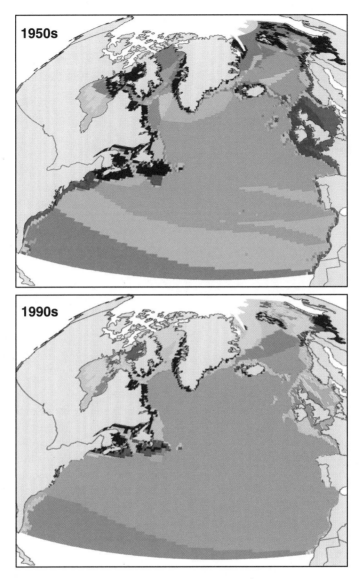

FIGURE 22. WHERE MARINE MAMMALS OBTAIN THEIR FOOD.
These maps show the pattern of average yearly food consumption for all marine mammals in the North Atlantic during the 1950s and the 1990s.

Red indicates high consumption rates of 4.5 and more metric tons per square kilometer per year. Note the decline in overall food consumption, due to the decline of overall marine mammal biomass from the 1950s to the 1990s, and the continued importance of New England/Eastern Canada and the North Sea and surrounding areas as major feeding grounds. Maps from Kaschner et al. (2001).

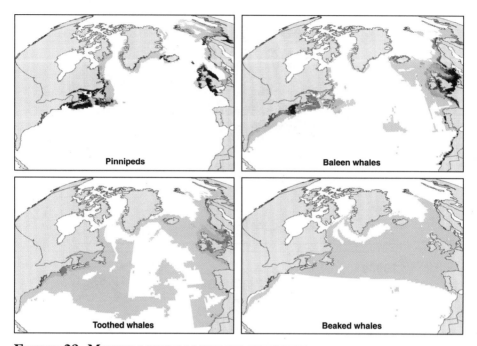

FIGURE 23. MARINE MAMMALS VERSUS FISHERIES.
Overlap between the prey of marine mammals and the catch of fisheries in the North Atlantic in the 1990s.

For pinnipeds (seals, sea lions, etc.) and baleen whales, there are areas of high overlap off New England and Eastern Canada, and in the North Sea and surrounding areas. The food eaten by pinnipeds and baleen whales reflects mainly the diet of the most abundant species, the minke whale (which feeds mainly on a variety of commercially exploited small pelagic and miscellaneous other fishes). Other groups show almost no overlap with fisheries. Among them are the toothed whales, whose food consumption pattern is dominated by that of the very large and relatively abundant sperm whales (which feed mainly on large deep sea squid that are not commercially exploited) and the oceanic beaked whales. Maps from Kaschner et al. (2001).

on the east coast of North America and in some European shelf waters. Not surprisingly, these hotspots coincide with areas of known conflict between fishers and marine mammals.

The World Wildlife Fund[38] proposed several mechanisms to improve the status of marine mammal populations worldwide: regional agreements for small whales and dolphins that may be beyond the purview of the International Whaling Commission; more marine protected areas, especially in critical breeding and nursing areas; large oceanic sanctuaries; encouragement of well-managed whale watching activities; reduction of bycatch, especially by enforcement of present regulations; and reductions in marine pollution. We agree.

The maps presented in Figure 23 may help in focusing these efforts on those areas where the dietary overlap between marine mammals and fisheries would yield high returns in terms of conservation. Moreover, these maps indicate that crude estimates of overall food consumption by marine mammals from the world ocean, recently bandied about in support of various "culling programs," misinform rather than inform the public and indeed, completely miss the complex geography underlying interactions between fisheries and marine mammals.

Report Card for an Impoverished Ocean

The marine food webs in which we have become enmeshed are very dynamic, often switching from one state to another in response to natural environmental change.[39]

Thus, the fish we take each year are being replaced at varying rates as the ecosystems respond to differing environmental conditions, as each year is always a little different from the previous one. And when we remove large chunks of a food web, much of its resilience must be lost and it will be unable to respond properly to the next change in the environment, unable to recover to its former state. It is not hard to visualize how such food webs, such ecosystems, can over time lose their integrity and become very different or "unhealthy"—if we accept that they were "healthy" in their previous unfished or pristine condition.[40] Taken as a whole, the world's oceans are all suffering from damaged ecosystems due to excessive fishing. The term "crisis" has been applied to the fisheries situation for at least a decade.[41]

An examination of the impact of North Atlantic Ocean fisheries can be seen, therefore, as an assessment of the "health" of its ecosys-

ecosystems. A single indicator of health status would be the ideal result. While no obvious candidate has appeared in the ecological literature, one of the best known ecologists, Eugene Odum, used the term "maturity" to distinguish between ecosystems at different stages of organization: fully mature on his scale corresponds largely to undisturbed, or "healthy." Some of the implications are that in more mature systems, all niches, or habitats, tend to be filled.[42] It also means that the "primary production," the new biomass produced by marine plants and the driving force behind all ocean life, is more fully used and recycled. Indeed, studies have since confirmed these features: indices of maturity in fished and unfished ecosystems are in agreement with Odum's theory.[43]

The notion of ecosystem health is still somewhat controversial, however, and we have chosen, therefore, to express the state of the North Atlantic using a metaphor based on something we all know: the "report card" that children so dread to bring to their worried parents.

A report card for the North Atlantic Ocean, based on the summary findings presented earlier, would show a series of poor grades in all factors related to the living resources, leading to the conclusion that the North Atlantic as a whole is indeed in bad condition. It would look like this:

Box 2.4

NAME: North Atlantic Ocean
CLASS: HEALTH STATUS

SUBJECTS	GRADE
Long-term Productivity of Fisheries	F
Economic efficiency of the fisheries	C-
Energy efficiency of the fisheries	D-
Ecosystem status	F
Effects of fisheries on marine mammals	D

CHAPTER 3

How Did We Get Here?

This chapter briefly explores how North Atlantic fisheries and their supporting ecosystems reached their current dismal state. There are no villains, least of all the fishers. Although it is true that the main factor in the decline of the North Atlantic is overfishing, the fishers are guilty only of trying to make the best living they can, taking advantage of the incentives they are offered. Some of the most perverse incentives,[1] e.g., encouraging fishers to discard bycatch, were inadvertently created by regulatory agencies overwhelmed by the magnitude of their mandates. These agencies are often understaffed and lacking the strong political support that is required for them to do their job. But just as importantly, as detailed in the previous chapter, the scientific information available to managers has typically been inadequate, leading to a misdiagnosis of the ocean and its problems. The result is a management regime at odds with biophysical reality.

A Conceptual Failure

In the introduction to this book, we noted an important difference between terrestrial agriculture and the oceans: Terrestrial agricultural systems tend to be fenced off open spaces, and in the sea they are not. Although few places on land are isolated from human impact altogether, in the sea, the distinction between human and wild disappears altogether. We cannot partition and cut up the ocean; we cannot divide the ocean between "ours" and "theirs." When we take fish from the ocean, we are not farmers harvesting the fields we have sown and tended; rather we are hunters, the top predators in the ecosystem, more efficient and more voracious than any other inhabitant of the marine food web.

Remarkably, however, we have historically made little distinction between terrestrial food production systems and marine ecosystems. In the U.S. for example, the oceans were until recently viewed as a means of combating world hunger. In 1966, the U.S. Congress funded two experimental plants to produce a powdered dietary supplement made from low-valued fish.[2] The powder was known under the less-than-appealing name "fish protein concentrate." Although the program failed, it was indicative of the mindset that the oceans constituted a plentiful supply of protein, and the only management problem was how best to exploit it. The program was part of a widely shared vision that we could literally farm the oceans—a vision which has, in a form, come to pass. We now have abundant, relatively inexpensive (compared to the cost of wild fish) fish protein, in the form of farm-raised salmon. Those salmon are, in a sense, processors of wild-caught forage fish protein, which is fed to the salmon who metabolize it into the special flavor and texture we value. Unfortunately, the salmon are not particularly efficient, typically converting only about one third of the forage fish protein they are given into salmon flesh.

It was common in the 1950s for governments to project massive increases in the levels of productivity of the oceans under scientific management. Although no more than 90 million tons of fish and shellfish have ever been harvested from the world's oceans, it was not unusual for scientists and agency officials to claim that 400–500 million tons might be caught.[3]

Ironically, the same agencies responsible for developing and administering these industrialized food production systems eventually became responsible for conserving those fish stocks, once it became clear that they were declining and falling far short of their projected potential. Nonetheless, the tools and frames of reference that led to the declines remained largely the same, even as the goals took a 180-degree turn. In the 1950s, at the height of the exploitation phase of fisheries management, scientists developed the concept of maximum sustainable yield, or MSY, described above. MSY was predicated on the idea that, up to a point, catching fish actually increased the amount that could be caught. The theory was not dissimilar to that justifying hunting of terrestrial species: culling herds makes more food available for the remaining members, thereby making them larger and healthier. Remarkably, the theory justified setting annual catch allowances at about half the population size of some fish species.

Of course, setting target catches in the vicinity of MSY did not take into account the environmental variability in the North Atlantic described in Chapter 1, and the kinds of changes in recruitment patterns that this induced. More importantly, the approach requires reasonably accurate estimates of actual population sizes. But like many estimates of the potential productivity of the oceans, many of the population estimates underlying MSY values were too high, if not wildly overoptimistic. Decision makers then tended to exploit any uncertainty in favor of exploitation. The result was that many fish populations were driven into depletion before their managers realized that they were even declining.[4]

Subsequent standards for allowable catch have become more conservative, and the science behind them has improved. In the 1970s, an attempt was made in the North Sea to account for the predator-prey interactions between major commercial species. This approach, called multispecies virtual population analysis (MSVPA), incorporates the main biological interaction (predation) of the target species with its predators and most of their prey. It has been used in the International Council for the Exploration of the Sea (ICES) area of the Northeast Atlantic and, as it depends on data from stomach contents analyses of the species concerned, has led to major studies on the diet composition of North Sea fishes. One study conducted in 1981 involved over 40,000 fish stomachs. Although potentially useful for long-term assessment of fish populations, MSVPA has never been used for setting long-term policies. Thus, the single-species approach has continued, albeit often informed by the improved MSVPA estimates of predation-induced natural mortality.[5] Unfortunately, because it considers only commercially important fishes, and because its data requirements are enormous, and for various technical reasons, MSVPA does not take us a long way toward ecosystem-scale assessments.

The single-species approach, now based on an extremely complicated version of its 1950s precursor's, still dominates North Atlantic fisheries management. While those stock assessments, based on national reporting, are of critical value to our assessment and provide a significant part of our database, they portray only one facet of the ocean, and it is not the facet most important to understanding the health of fisheries and the steps that must be taken to manage it effectively.

When fishers extract fish from an area of the sea, they affect the properties—the balance or dynamics and even the physical structure—of the ecosystem in that area to some extent. As our analysis shows, the state of the North Atlantic is much more dismal than the assessment of any one fish stock could indicate, and the reason is that until now no assessment has been made of the effect the removal of multiple predators has on the structure and productivity of the ecosystem as a whole. Management thus takes no account of these effects.

Distorted Economics

A legitimate question to ask upon reaching this chapter is how North Atlantic fisheries, which are run as commercial enterprises, can continue to operate, given the decline of their resource base. One should expect that the continuously rising cost of catching fish, and their continuing decrease in abundance, would bankrupt the sector, because at some stage consumers should become unwilling to pay the escalating market prices. So far, it has not worked that way. People are still willing to buy, although the price of fish has been rising much faster than the average price of consumer goods, as shown above.

One of the major reasons for the continuing buoyancy of over-fished fisheries is that they are subsidized by taxpayers' money (see Box 3.1).[6] Several international organizations involved in looking at the economic impact[7] of government subsidies have concluded that they are clearly responsible for over-investment in gear and vessels working in fisheries that could not otherwise support it. A prominent example is the expansion of the U.S. fishing fleet in the late 1970s and 1980s following declaration of an exclusive economic zone. Two government-funded programs—The Fishermen's Capital Construction Fund and the American Fisheries Promotion Act—led to increases in fleet size well beyond what the resources could bear. Although it is often overlooked, the income tax credit (ITC) also played a large role in attracting capital into fleets. The ITC allowed people to shelter money and take losses in investments to which they had no other link. So, doctors and lawyers invested in boats that they never set foot on, adding to what economists refer to as overcapitalization.[8] The fisheries have a far greater investment in capital equipment than can be justified by the returns they generate. While the subsidies have enabled fishers to be ever more effective in catching fish and

---

BOX 3.1 SUBSIDY

Any government program that potentially allows a firm to increase its profits, directly or indirectly, beyond what they would have been in the absence of the government program. Subsidies, by effectively reducing costs, increase the amount of fishing that can undertaken before reaching the break-even point, when costs and returns become equal (and all economic rent is dissipated). Similarly, price increases, by increasing gross returns, increase the fishing effort that can be deployed before all the economic rent is dissipated.

maintaining their operations, they are also a primary cause of the massive overfishing of the North Atlantic.

On a worldwide basis, the total annual cost of fisheries subsidies is enormous, possibly between US$15 billion and $20 billion, or even as much as $50 billion per year,[9] while for the North Atlantic alone, total annual subsidies (including costs of research and management, but not price support) were estimated as US$2.5 billion per year for the late 1990s (Figure 24). Among individual countries, subsidies, at one or more points in the chain from building vessels to catching and marketing fish, ranged from 5% to nearly 50% of the landed value of the catch. The average rate of fishing industry subsidies for European Union countries was 16% of landed value, for other European countries 10%, Canada 25%, and the U.S. 24%.[10]

The effects of subsidies are numerous. In addition to increasing fishing capacity and effort, they affect distribution of income—those who receive them are better off, without making any extra economic contribution to society, than those who do not—and they have an impact on resource management and sustainability. There is another distributional dimension to these subsidies: the general public pays for them through taxation, but is rewarded with cheaper fish only in the short term. In the longer term, subsidies lead to resource scarcity and, ultimately, to higher prices.

For the North Atlantic, subsidies totalling US$367 million for fisheries infrastructure, investment, modernization, and tax exemptions, contribute directly to the overcapacity of the fishing fleet and thus must be deemed negative.[11] Of the remainder, subsidies for decommissioning (buy-back of vessels) and for access to other countries' waters, which were 23% of the total subsidies in 1997, are considered by many observers to be positive in terms of resource conservation. This is true, however, only for the area from which the surplus vessels originate; for the receiving countries, the surplus effort resulting from this can have a devastating impact on already overfished resources, as is happening, for example, in northwest Africa with vessels from the European Union.[12]

At present, several North Atlantic countries, including Canada, Spain, the United Kingdom, and the U.S., are carrying out

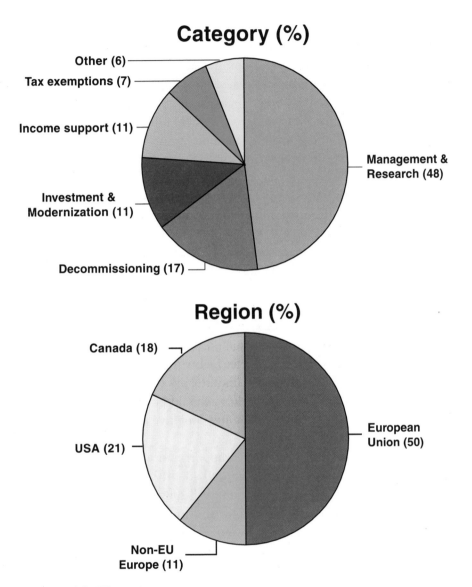

FIGURE 24. WHO GIVES SUBSIDIES, AND FOR WHAT.
Breakdown of the estimated 2.5 billion U.S.$ in annual fisheries subsidies, by country/region of the North Atlantic, and by type of subsidy.

The numbers presented here are adapted from an OECD study, analyzed and expanded upon by Munro and Sumaila (2001).

decommissioning programs in which the government pays fishers to take their vessels out of a fishery.[13] This, in effect, subsidizes those remaining in the fishery, who thereby face reduced competition. While it might seem obvious that such programs would be positive in their impact on conservation of ecosystems, this is open to serious question. In order for them to have a positive effect, two conditions must be met. First, there must be no tendency for fishing capacity to "seep back" into the fisheries over time. Second, decommissioning programs must be unanticipated by the industry. Anticipated decommissioning payments can spur investment in fleet capacity. When anticipated, they can, perversely, serve as collateral for the financing of new vessels.

The conditions for successful decommissioning are seldom met. The case of the European Union was mentioned earlier,[14] where recent efforts to decrease fishing capacity have been a general failure.[15] Indeed, decommissioning programs have negative economic outcomes in anything less than a perfectly managed fishery where fishers have zero anticipation of such a program. It can be shown that in any other case, that is, in the real world, there will be large economic waste compared to not undertaking the program at all.[16] Decommissioning subsidies, therefore, should be viewed with deep suspicion, especially because decommissioned vessels often end up being exported to other fisheries.

Subsidies in the form of income support and unemployment insurance, where linked with fishing activities (that is, where they depend to some extent on the amount of fishing undertaken), allow fishers to remain in the industry and so maintain their level of potential effort or capacity beyond what is economically viable. This raises the price of fish over time and so these subsidies are negative from society's viewpoint. For the North Atlantic, subsidies of this type in 1997, directly related to fishing activities, were almost all in Canada.[17] Under the present conditions of almost universal overfishing, maintaining fishing capacity has its own, detrimental impact on marine ecosystems, as described in previous chapters.

For the North Atlantic as a whole, at least 50% of the $2.5 billion in annual subsidies provided to the fishing industry in the mid to late

1990s had a negative financial impact on society and undermined the sustainable use of the ocean's living resources.

The other 50%, devoted largely to funding research, management, surveillance, and enforcement, should be neutral or positive in its effects on sustainable use of the ocean's resources, assuming these activities were effective. We know from the shrinking and altered nature of the ocean's ecosystems that it is likely not the case.

Economists argue that subsidies are particularly damaging under conditions of open access and in unmanaged common-property fisheries. This is not disputed. However, subsidies can be seriously damaging even if the common-property aspect of a fishery is removed, for example, by establishing an effective rights-based fishery, as with a quota system. Even where property rights have been given to encourage stewardship, the subsidies can result in "asset stripping," a term that arose in the heyday of corporate raiders, when the hostile new owner would sell off the company's assets, reducing it to an empty shell. Under certain circumstances, the introduction of subsidies to an apparently well-managed fishery (in which, perhaps, fishers' returns were unreasonably low) can give the new owners the incentive to fish in the most profitable manner possible, even if it is destructive, and then move on to greener pastures.[18]

Small-scale vs. large-scale fisheries

Among the reasons given for subsidizing fishing is that larger firms with better equipment in larger-scale vessels are more efficient than numerous, less well-equipped, smaller-scale fishing operations. Although initially subsidies of fleets were intended merely to spur investment, a number of large-scale fleets have now become dependent on subsidies, bringing into question the often-aired assumption that they are more economically efficient than small-scale operations. At the same time, the consolidation of the fishing industry into ever smaller numbers of hands and larger vessels causes serious economic and social dislocations in fishing-dependent communities.

In view of this, we compared the economic profitability of the small- and large-scale sectors, in order to determine the best "economy-of-scale" and to objectively evaluate the benefits of large-scale fishing operations against their costs. To enable comparison between areas with different fleet structures, we developed a new definition of small and large scale that takes into account the diversity of gear types and vessel sizes used in various countries. Essentially, for each country, we ranked the commonly used fishing gear and vessels from large to small (in terms of the typical catch per year of one such gear/vessel combination), and computed, for a given reference year, the mean annual catch of each gear/vessel combination. Then, for each country, starting from the smallest gear/vessel combination (e.g., handline/no vessel) to the largest (e.g., trawl net/trawler), we identified the group of gear/vessel combinations that contributed half of the catch value of that country, thus identifying these as constituting the small-scale sector (the large-scale sector, obviously, consisted of the other half of the gear/vessel combinations). This flexible definition, while allowing for comparisons between the small- and large-scale sectors of different countries, has the advantage of allowing for differences in the absolute sizes of the gear and vessels considered "small" or "large" in these countries (Figure 25). [19]

In Norway, where about half the catch is processed for fish meal and fish oil, small-scale fisheries employ five times as many people as large-scale fisheries and generate five times as many jobs per unit of landed value. Small-scale fisheries also achieve 150% better value for their catch than do large-scale fisheries, because 60% of the latter catch is low-value fish for processing into fish meal and oil, while only 15% in the small-scale sector catch is so used. Similar results are obtained for other North Atlantic countries.

Large- and small-scale fishing fleets could both improve their economic efficiency by cooperating. An analysis of the fisheries in the Gulf of Maine and adjacent Georges Bank indicate that net returns on investments would be maximized by reducing the amount of fishing by the small-scale fleet by about one-tenth, and the large-scale fleet by one-third.[20]

Other things being equal, this redistribution of effort would add about US$110 million in extra profit to the two sectors in a fishery that was worth perhaps $650 million at the time of the analysis. The

Fishery benefits	Small-scale	Large-scale
Number of fishers	18,600	3,200
Number of vessels	13,000	300
Annual catch (thousand metric tons)	860	2,000
Annual catch (thousand metric tons) for human consumption	724	816
Landed value (million US $)	711	674
Fishers employed for each $1 million landed value	26	5
Annual catch (thousand metric tons) for reduction to fishmeal and oil	130	1,200
Total fuel consumed (million litres)	350	300

FIGURE 25. SMALL IS BEAUTIFUL?

Comparisons of the small- vs. large-scale subsectors in Norwegian fisheries, using data for 1998.

Small-scale fisheries, in the North Atlantic and elsewhere, often share attributes required for sustainability, especially when the fishers themselves are involved in decision-making. However, they are equally often pushed aside by large-scale fisheries, which tend to have better access to global markets and to political support. Note that the similarity between the landed values of these two subsectors results from their definition. From Sumaila et al. (2001).

Gulf of Maine/Georges Bank fisheries are a microcosm of the North Atlantic, for which the total losses for all fisheries scale up to an enormous figure. By re-apportioning part of the catch to small-scale fishers, there would be gains in efficiency, meaning less cost and less waste. Such a re-apportionment would also mean gains in employment and a wider societal spread of the increased profits, as implied by Figure 25.

Present vs. future generations

A larger reason why we find ourselves with an impoverished ocean is because past generations did not care about us and discounted what appeared to them as future benefits to be gained from exploiting the North Atlantic. Our generation is no better. We also discount the benefits future generations may obtain by fishing in the North Atlantic. In effect, our current policymaking framework encourages us to take fish from the oceans today, rather than leave them in the ocean to benefit future generations. A factor in policymaking known as the discount rate essentially calculates that the value of an unexploited resource declines over time (by some percentage for each year), and eventually approaches zero, at least as far as our generation is concerned. Thus, when looking at the costs and benefits of a policy, the ledger sheets are weighted against preserving fisheries and their supportive ecosystems for future generations.

Neoclassical economists justify their failure to consider future generations in part with the claim that our descendents will have found substitutes for depleted resources. There is some validity to this claim. For example, copper prices have not risen as rapidly as might have been expected, in part due to the development of products such as PVC piping and fiber optic cables, which have reduced demand for copper pipes and electrical cables. But copper does not reproduce itself, nor does it play a key role in maintaining the "ecosystems" from which it is mined. Low-trophic-level farmed fishes like tilapia may be an adequate substitute for wild, high-trophic level fish on the dinner plate, but they cannot replace the roles that the wild fish played in the ecosystems from which they were extracted.

However, since fisheries policy analyses discount future fishery benefits, the result is that cost-benefit analyses are skewed toward exploitation of the resource by the current generation. This puts the burden of proof in policy debates on those who would argue for restoration.

Ineffective Governance

We have devised systems of management wholly inadequate to the task of overcoming these formidable obstacles. We have not used the regulatory tools and developed the right institutions for managing fisheries effectively for their long-term sustenance. At the national level, governments find their ability to set and enforce fishing regulations compromised both by political interests as well as overlapping jurisdictions. For example, in the United States, the National Marine Fisheries Services can offer advice and even give mandates, but these are then acted on by individual states, which generally control their three-mile coastal zones, and by eight regional management councils that set catch limits and design stock recovery plans. In Europe, a European Union Commission provides similar guidance to a Council of Ministers, which is then left to set policies, usually formulated to benefit the short-term interests of the industrial fleets of member countries.

Whether successful or not, as presently operating fisheries management is very costly, and must be viewed as a form of subsidy to the industry, to the extent that management is aimed at enabling the orderly and maximum exploitation of fisheries resources (and it usually is, though fishers may nonetheless resent it). Of course, to the extent that management agencies protect the resources as a public trust, such costs to taxpayers are entirely appropriate.

Management entails a large infrastructure of research, administrative, monitoring, and regulatory entities, which are usually paid for by the taxpayers.[21] In OECD countries, which include those around the North Atlantic and some others, management costs in 1997 were a little over one-third of all government spending related to fisheries. The research component accounted for 8%, management services

16%, and enforcement 12%. The total costs amounted to 6% of landed value of the fisheries and ranged from less than US$3 to more than $400 per metric ton of fish landed.[22]

Generally, management costs ranged from about 3% of landed value in Iceland to about 10% in Norway for 1989–1996.[23] However, they were much higher in Newfoundland, Canada, ranging between 18% and 28% of the catch value from 1989 to 1998. Overall, annual expenditure of the Council of Fisheries and Oceans in the Newfoundland region in that period ranged from one-third to more than one-and-one-half times the catch value.[24]

The costs of some government functions rise with the complexity of the fisheries, their supporting industries, and the regulatory tools used. Present fisheries management tends to rely more and more on fine-scale adjustments of catch limits that require more and more effort to validate and monitor in the field. Expensive research surveys are also needed which require larger vessels as the fisheries have moved farther offshore, as well as observer schemes and even satellite tracking of the position and speed of fishing vessels. Even wealthy countries such as the U.S. devote inadequate funds to do the job. Less well-to-do countries simply do not have the means to manage their fisheries. Ashore, market regulations have had to become more sophisticated to ensure, for example, the correct naming of the fish species used in various products. Further increases in such management-associated costs will inevitably occur as more popular fish become scarcer and the temptation to use cheaper substitutes becomes stronger.

Thus, not only have most management measures proved unable to control the rapid decline of almost all major commercial fisheries, they have done so at great cost. A closer look at the policies and institutions that have led to this situation follows.

Local and national governance

Around the North Atlantic as elsewhere, fisheries are managed primarily at the national level, and typically are controlled by governments. In countries with several levels of government, federal or national governments and state or provincial governments have generally shared responsibility for legislation and its enforcement (except for Canada, where essentially all marine fisheries are run by the federal government). Where both levels of government are present, the national government often manages the broad 200-mile exclusive economic zone (EEZ), while leaving the individual fisheries for the state/provincial government to manage. National governments also get involved where fisheries expand into international waters and thus require international arrangements (see below).

A strong, direct government role creates a form of top-down fisheries management that makes enforcement of regulations and collection of reliable data exceedingly difficult, due to resistance in the affected fisher communities. This is one reason, for example, why catches remain undeclared, and gear are deployed which are not legally sanctioned. Just as in the case of control of vehicular traffic by police, control by fisheries officers cannot fully replace voluntary compliance by the majority.

National government attempts to constrain fishing effort have involved a variety of approaches,[25] including financial incentives (depreciation schedules, allowances, tax exemptions), penalties (fines, seizures), taxes and charges (access charges, license fees), property rights (licenses, quotas), and protected/restricted areas (closed seasons, moratoriums, closed areas).

There have been some success stories. For example, in Norway, the use of larger mesh sizes on driftnets is credited with preserving juvenile cod, albeit at quite low levels of abundance. Nonetheless, experience, often bitter, has shown most of these measures to be flawed. Fishers almost always find ways to increase their capacity to catch through legal loopholes or illegal means. A good example of a loophole is the consequence of the decision in the late 1970s to split Canadian fisheries into two groups, inshore and offshore, based on size of vessel. The intention was, among other things, to keep large vessels out of the fishing grounds of the small-scale inshore fisheries. The inshore fishers, whose vessels were mainly quite small, virtually en masse ordered larger vessels, an inch shorter than the prescribed upper limit of the inshore category. These vessels were big enough to exploit both offshore and inshore grounds, greatly increasing fishing effort in both areas.[26]

The problems created by top-down management have led many nations and the EU to pursue local co-management of fish stocks. Co-management may mean anything from local self-management to agreements between fishing industry organizations and national governments. The important characteristic of these arrangements is the inclusion of fishers in the decision-making processes, e.g., in the setting of total allowable catches (TACs), the determination of fishing seasons, decisions to limit or close fisheries, etc. The theory is that by including the affected user groups in making decisions, the regulations will be designed with their interests in mind, while pursuing conservation goals at the same time. This is likely to result in better compliance and reduced enforcement costs.

In fact, in several instances, in the New England lobster fishery and the Danish Matjes herring fishery, for example, most participants and observers say that co-management has been successful. In the lobster fishery, this is due primarily to a set of factors that are not present in most other fisheries, including cultural systems of entry and enforcement that predate the implementation of the co-management regime, as well as the behavior of the lobsters themselves, which allow for greater territoriality among fishers. The Danish Matjes fishery, while also deemed a success, is threatened by the market expansion of the Norwegian Matjes fishery and by the demands of the Danish trawling industry for entry into the fishery. These threats point to an overarching challenge for many co-management regimes: the difficulty of isolating a fishery both geographically and from international market forces.[27]

The U.S. has also pursued a type of co-management that points to a different kind of difficulty. Following the Magnuson-Stevens Fisheries Conservation and Management Act of 1976, eight Fishery Management Councils were created that were meant to allow public participation in the process of managing fisheries, previously the sole prerogative of government agencies at state and federal levels.[28] Unfortunately, the "non-agency" members of the councils, nominated by the governors of the states covered by these councils, turned out to consist overwhelmingly of representatives from the commercial and recreational fishing sectors. Magnuson-Stevens was seriously flawed as a mechanism for fisheries management, as it fueled fleet expansion and contained significant loopholes in setting allowable catches. It was reformed in 1996 but, in spite of several amendments aimed at a precautionary approach to management and eliminating loopholes in the original act, the fact remains that the key management decisions are made by the Fishery Management Councils, which show a marked tendency to put resource conservation on the back burner, with results that could have been predicted, and which have now been described in saddening detail.[29]

For example, despite a series of ultimatums from the U.S. National Marine Fisheries Service (NMFS) beginning in 1992, the New England Fishery Management Council did not adopt a management plan for severely overfished monkfish populations until 1999. In that year, fishers landed 25,000 metric tons of monkfish—up 4,000 metric tons from the catch in 1992, but below the 1997 peak of 28,000 metric tons. In 2000, NMFS scientists concluded that despite restrictions, monkfish were still overfished. Such episodes are the norm, unfortunately, and although the Fishery Management Councils may be better than some alternatives, it would be hard to argue that as presently constituted they have not contributed to the decline of major fisheries.[30]

Politics often dilutes the best management schemes at the national level as well as at the intergovernmental level. In Europe, for example, where policy recommendations are made by the European Commission to the Council of Ministers, which then sets policies for EU member states, advice on how to reduce fishing effort in European waters was proposed in 2000 by the commission. However, that advice was distorted by the Council of Ministers, which minimized the commission's proposed reductions, to the extent that the resulting initiative not only failed to reduce capacity, but led to actual increases in fishing capacity. Despite that, European fishers were only able to catch 60,000 of the 80,000 metric ton cod quota for 2000—

which itself was a reduction of 40 percent from the previous year—
and the commission has recommended a decrease of three quarters
of that quota for 2001.[31]

On May 28, 2002, the commission announced yet another initiative to reform the European Union's Common Fisheries Policy and
reduce fleet capacity, but six of the fifteen members of the Council
of Ministers have announced that they will not accept the proposed
reform.

International arrangements

National governments are limited, of course, by the fact that fish,
particularly large pelagic ones, do not respect exclusive economic
zone (EEZ) boundaries. Moreover, regulations vary from nation to
nation, and enforcement among nations is often difficult. Fisheries
cooperation in the North Sea[32] area began with a convention in 1882
to reduce conflict between the different nations fishing there. Organized fisheries research was still unknown. However, after much
preparatory effort, the International Council for the Exploration of
the Sea (ICES) was established and first met in 1902, for research in
the northeastern Atlantic, including the North Sea, the Baltic and ad-
joining seas. The council has continued up to the present time to provide advice on fisheries in these areas through its various committees.
Since World War II, many agreements relating to the North Atlantic
Ocean have been made and revised or superseded as conditions
changed. About 15 current instruments deal directly with fisheries
matters (see Table 3.1). Of these, the major instruments in the North
Atlantic include the International Commission for the North West
Atlantic Fisheries (ICNAF), established under the terms of the International Fisheries Convention in 1946 (the overlapping interests
of ICES, ICNAF and the Intergovernmental Oceanographic Commission [founded in 1961] resulted in an Inter-Agency Coordinating
Group for North Atlantic Oceanography that first met in 1968); and
the North East Atlantic Fisheries Commission (NEAFC), for which
ICES provides scientific advice on populations.[33]

Most agreements cover specific areas rather than species (Figure

TABLE 3.1. EXISTING INTERNATIONAL INSTRUMENTS DEALING WITH FISHERIES OR RELATED TOPICS IN THE NORTH ATLANTIC OCEAN

INSTRUMENT	NOTES
Conservation and Management of Straddling Fish Stocks and Highly Migratory Fish Stocks (**UNCLOS Fish Stocks**)	A component of the UN Convention on Law of the Sea (UNCLOS); not in force, waiting for 3 signatures
Agreement to promote Compliance with International Conservation and Management Measures by Fishing Vessels on the High Seas (**UNCLOS Compliance Agreement**)	A component of the UN Convention on Law of the Sea (UNCLOS); not in force at the time of writing, as only 12 signatures are available, out of the required 25.
Convention for the International Council for the Exploration of the Sea (**ICES**)	Provides scientific advice to regional fisheries bodies in the NE Atlantic
International Convention for the Conservation of Atlantic Tunas (**ICCAT**)	Manages tuna and billfish in the entire Atlantic area
Convention on Biological Diversity (**CBD**)	The CBD includes the Jakarta Mandate which is addressing the issues of marine biodiversity in marine environments
Convention for the Conservation of Salmon in the North Atlantic Ocean (**NASCO**)	Management of salmonid fisheries throughout the North Atlantic
Convention for the Protection of the Marine Environment of the North East Atlantic (**OSPAR**)	Addressing marine pollution in the NE Atlantic, in particular the North Sea
Convention on Future Multilateral Cooperation in the Northwest Atlantic Fisheries (**NAFO**)	Management of mostly groundfish in the NW Atlantic area
Convention on Future Multilateral Co-operation in Northeast Atlantic Fisheries (**NEAFC**)	Management of mostly pelagic fish in the NE Atlantic
Common Fisheries Policy (**CFP**)	Management of over 100 species of fish within the waters of EU member countries
Agreement Concerning Certain Aspects of Cooperation in the Area of Fisheries which also includes the following three agreements	This agreement and the following 3 address management of the major fisheries in the Barents Sea that are not encompassed in NEAFC or CFP

continues

TABLE 3.1. (*Continued*)

INSTRUMENT	NOTES
Agreement Concerning Cooperation in the Field of Fisheries Between Norway and the USSR (1975)	–
Agreement Concerning Mutual Fisheries Relations Between Norway and the USSR (1976)	–
The Grey Zone Agreement Between Norway and the USSR (1978)	–
Agreed Record of Conclusions of Fisheries Consultations on the Management of the Norwegian Spring Spawning Herring (Atlanto-Scandian Herring) Population in the Northeast Atlantic for 1997 (Including Supplementary Agreements)	Management of herring population fished primarily in the Norwegian Sea
Negotiations on Allocating the Capelin Stock Between Norway, Iceland, and Greenland	Management of capelin population fished primarily in the Jan Mayen area
International Convention for the Regulation of Whaling (**IWC**)	See text
Agreement on the Conservation of Small Cetaceans of the Baltic and North Seas (**ASCOBANS**)	–

26). Exceptions are the Convention for the Conservation of Salmon in the North Atlantic Ocean (NASCO) and the International Commission for Conservation of Atlantic Tunas (ICCAT), the implementing body of a convention to protect these tunas, signed in 1966.

The functioning and general ineffectiveness of these instruments is examined below, with respect to compliance and enforcement issues as well as the scale of fisheries. However, more generally, the effectiveness of these instruments is compromised by the fact that not all fishing countries may be signatories to a given agreement. This not only allows the fishing fleets to overexploit a stock on the high seas, it allows vessels originating in signatory countries to fly "flags of convenience," by registering their boats to nonsignatory nations.

A further challenge is that even among those that are signatories, lack of agreement of one country to a provision, such as an annual catch allowance, often means that no country must abide by it. The allocation of catches is often contentious, and this leads to the setting of TACs at higher levels than the stock assessment might warrant.

Finally, the instruments themselves provide for no enforcement measures other than political pressure and nominal economic sanctions against offending countries. Such measures are of varying effectiveness, particularly where the offending country itself lacks the means to enforce compliance among its fishers.

Fisheries compliance with international instruments

Many international instruments, some voluntary, some binding to contracting countries, have been applied to fisheries that occur in the high seas or cross national borders in the North Atlantic. There have been at least 43 instruments of a multilateral or international nature involving three or more countries, as well as numerous bilateral agreements. Of the total, nine have been overtaken by political or other developments and have been superseded; 19 have only limited relevance to fisheries; and 15 are directed mainly at fisheries. There is enough information on catches and fishing activities to rate, using a number of criteria,[34] the relative compliance of contracting countries with the 15 instruments aimed mainly at fisheries and listed in Table 3.1. The parts of the North Atlantic covered by the major instruments[35] are shown in Figure 26.

The main issues affecting compliance with international instruments are the costs of monitoring; the balance of political pressure versus scientific advice; the quantity and quality of information; and the capacity of the member countries to implement fisheries management measures.

In general, there is a north-south gradient in formal compliance with such instruments, the scores being highest in Nordic countries and Canada. These countries are active in fisheries management; enforce the use of logbooks; maintain observer schemes; and generally fish within quotas. The United States is less compliant than the Nordic countries because of its lack of formal commitment to UNCLOS, which is partly offset by its active involvement in fisheries agreements related to UNCLOS, as well as in ICCAT and the IWC.

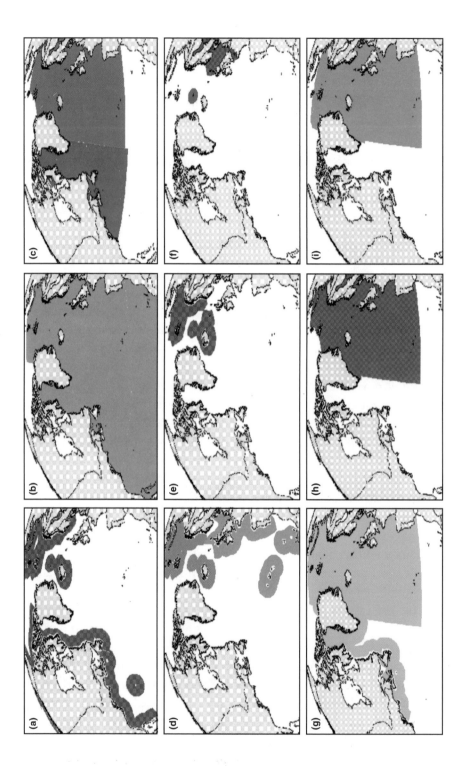

FIGURE 26. INTERNATIONAL FISHERIES MANAGEMENT.

Parts of the North Atlantic covered by various international instruments devoted to fisheries management or environmental protection.

The instruments included are: (a) "Fish Stocks Agreement"; (b) International Convention on the Conservation of Atlantic Tuna (ICCAT); (c) Conventions on Future Multilateral Cooperation in the North East Atlantic (purple) and the North West Atlantic (red); (d) Common Fisheries Policy (blue) and Agreement Concerning Certain Aspects of Cooperation in the Area of Fisheries (purple); (e) Norwegian Spring Spawning Herring Stocks Agreement; (f) Capelin fishery in the Jan Mayen area (red) and Agreement on the Conservation of Small Cetaceans in the Baltic and North Seas (blue); (g) Convention for the Conservation of Salmon in the North Atlantic; (h) Convention for the Protection of the Marine Environment of the North East Atlantic (OSPAR); (i) International Council for the Exploration of the Sea (ICES) Convention. Maps from Alder et al. (2001).

Of the other North Atlantic countries, France, Spain, and Morocco are lowest in compliance with international instruments. These countries are unwilling to provide information, and if provided, it is often not reliable. Their vessels are often caught taking undersized fish or exceeding quotas, and logbook and observer programs are present in only a few of their fisheries.

In terms of formal compliance, the most effective instruments in the North Atlantic were found to be:

- The convention of ICCAT, the provisions of which are binding on members who otherwise face economic sanctions. There are at present sanctions against Belize and Honduras, which are conveniently small noncontracting parties, for fishing in contravention of ICCAT recommendations.
- The Fishing Cooperation Agreement, which deals with major fish species in the Barents Sea using the precautionary approach and advice from ICES, has made cuts in total allowable catches in order to maintain fish population sizes with quotas that are binding.
- NAFO, which has in recent years set total allowable catches for the main populations and has a comprehensive logbook program and observer scheme.

Other instruments with which there is relatively high compliance are:

- The North Atlantic Salmon Conservation Organization (NASCO), whose objective "to contribute through consultation and cooperation to the conservation, restoration, enhancement, and rational management of [those] salmon stocks . . . which migrate beyond areas of fisheries jurisdiction of coastal States of the Atlantic Ocean north of 36°N latitude."[36]
- The Convention on Future Multilateral Co-operation in Northeast Atlantic Fisheries (NEAFC), devoted to the management of pelagic and other fish in the northeast Atlantic.[37]
- The International Whaling Commission (IWC), which regulates whaling on a global scale, and whose North Atlantic members[38] adhere to its provisions, apart from Norway, which objects to the ban on commercial whaling and maintains a coastal whaling fleet.

- The Agreement on the Conservation of Small Cetaceans of the Baltic and North Seas (ASCOBANs), which is attempting to lower the dolphin bycatch rate in a small section of the northeast Atlantic, with active participation by a number of European countries.
- The Convention for the Protection of the Marine Environment of the North-East Atlantic (OSPAR), which focuses on pollution, but is also concerned with fish populations, for which it liaises with ICES, and whose recommendations are binding, requiring states to notify the secretariat on their implementation.

Generally, compliance by countries with the terms of instruments dealing with pollution and conservation of marine mammals is far better than that with fisheries instruments, possibly due to greater public awareness of pollution issues and, in the case of marine mammals, the "charismatic megafauna" phenomenon.

Ultimately, however, those instruments that most needed to be effective to protect ecosystems and high-trophic level fish have been shown to be ineffective. ICNAF (succeeded by NAFO) was unable to prevent the build-up of distant-water fleets from Europe and Asia that gathered in the Northwest Atlantic in the 1960s to participate in the then-lucrative groundfish fishery, which led to its first collapse shortly thereafter. Fish populations covered by the EU Common Fisheries Policy remain nearly all overfished. NASCO oversees the declining salmon fisheries. In each case, the problem was not necessarily in the design of the instrument per se, but in the inability to create international institutions capable of ensuring compliance.

What emerges from this overview of international fisheries instruments is that the countries with highest compliance ratings are also those whose fisheries rank highest on various sustainability scales.[39] Nevertheless, high compliance and high scores do not necessarily mean that the management measures themselves are adequate, as the dismal history of North Atlantic fisheries illustrate. Considering the number of instruments being used to protect the resources, 15 directly and a slightly larger number indirectly aimed at the fisheries, one can only express dismay at their results to date (see Figure 27).

As might be seen, each of the stocks or species covered by the

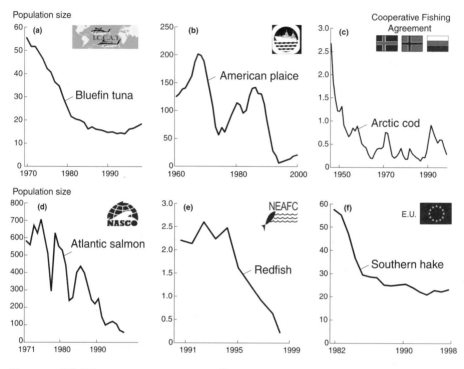

FIGURE 27. WHAT DOES IT MATTER?
The major types of fish under the care of the six international instruments above are all in a state of decline.

Population sizes of representative, formerly abundant species, expressed as adult biomass in million metric tons except as indicated; (a) Western Atlantic bluefin tuna, managed under the International Commission for the Conservation of Atlantic Tuna (ICCAT); (b) American plaice in Division 3LNO of the North Atlantic Fisheries Organization (NAFO); (c) Arctic cod, managed under a "Cooperative Fishing Agreement" between Iceland, Norway and Russia; (d) Atlantic salmon, in thousands of individuals, in areas covered by the North Atlantic Salmon Conservation Organization (NASCO); (e) Total biomass of Atlantic red fish, managed under the North East Atlantic Fisheries Commission (NEAFC); (f) Southern hake in Divisions VII & IXa of the International Council for the Exploration of the Sea (ICES), and managed under the Common Fisheries Policy of the European Union. Composite graph from Alder et al. (2001).

instrument in question, or which can serve as representative of the several stocks or species within that instrument's mandate, has been declining over the last few decades, a fact which, by itself, documents the failure of these instruments "on the ground." This reinforces the point made earlier that North Atlantic fisheries resources have not been managed sustainably, despite a wide range of instruments having been set up for that very purpose.

Institutions and Equity

Clearly, given the results presented above, the present institutions[40] that govern North Atlantic fisheries are far from efficient. This is due in some cases to their not having enough money and personnel. But they also suffer from a lack of political will among elected leaders to take the sometimes drastic steps needed to reform the fisheries, and an accompanying overabundance of political will to deny the future consequences of inaction.

There is a good case for decentralizing control of many fisheries. Indeed, turning away from international institutions to more local control of marine resources where appropriate (and there are many caveats) could greatly improve the attitude and level of responsibility of users of marine resources, both those who extract fish and those who only want to admire them. However, this would involve reining in the big, especially transnational, fishing companies whose size makes them difficult even for central governments to control.

Given the vagaries of the environment and political and social systems that can affect fisheries, governance institutions need to be flexible enough to cope with new eventualities. As in the evolution of animals and social systems, the more rigid their constitution, the less they can cope with new eventualities.[41] Current institutions in control of fisheries do not adequately include elements of local management where possible, or a large degree of social analysis, or a larger voice than in the past for the fishers themselves.[42] Nor do they bring to bear the voice of the nongovernmental organization (NGO) community—the only institution allowing for accountability at the global level.

As our perspective evolves from a fishing to an ecosystem perspective, we must allow the voices of the nonfishing communities to be heard. This need should be recognized early, before decentralization of fisheries management occurs, so that all users of the marine ecosystems, not just industrial or other fishers, take part in the ocean's stewardship. Of course, a key group of users that we must consider is future generations—our children and grandchildren. For them, preservation of the status quo is not enough. Restoration of past levels of abundance is required.

This is what we propose in the next chapter.

CHAPTER 4

What to Do?

The fact that the world community has so far failed to rec-
ognize the seriousness of the deterioration that has taken
place is . . . a major obstacle to change.
—Independent World Commission on the Oceans (1998).

We have only two basic choices. One is that we decide to treat the ocean as incurable. This is what we will be doing if we allow the status quo—a multibillion dollar commercial machinery extracting the last valuable scraps from the ecosystems it is destroying—to continue. We might go even further and cut short the present misery by withdrawing existing regulations—a frequent demand of some members of the fishing industry.[1] We could let the ocean's fisheries decline even faster. Eventual collapse of more and more fisheries is clearly indicated in either approach, the difference being only one of time. Clearly, if this should happen, we would be even less capable than at present of meeting the ever-increasing demand for fish.[2]

The other choice is that we seek to improve on the report card we presented in Chapter 2. This clearly requires more than attempts to improve upon any one part of the ocean or any one aspect of the ocean's health. In fact, the most important implication of the research and analysis presented in this book is that we cannot re-

research and analysis presented in this book is that we cannot re-store fish stocks without restoring the ecosystems that support them.

Ecosystem restoration implies ecosystem-based management, which itself implies going beyond the species-by-species, fishery-by-fishery approach presently used.[3] The Magnuson-Stevens Act in the U.S. includes ecological considerations in setting fishery manage-ment plans, requiring the identification of essential fish habitats and ways of minimizing the impact of fishing gear on fish habitats. How-ever, the term "ecosystem-based fisheries management" as used here and by various other authors and organizations retains the flavor of "fisheries management."[4]

Nowhere has this been more evident than in several recent pro-grams in which thousands of marine mammals have been "culled" (i.e., killed) to increase fishery yields in different parts of the world.[5] A rationale for culling is given in a recent report of the United Na-tions Environment Programme.[6] For an environmental program, the tone is decidedly fishery-centric: "Concern has been expressed over the quantities of fish consumed by seals, dolphins and other marine mammals, and the possibility that they are affecting the size and availability of the fish stocks and thus the viability of fisheries." Further, as the industry "attempts to maintain catches from gener-ally dwindling fish stocks, marine mammals are liable to be per-ceived increasingly as competitors to fisheries."[7]

The annual total food fish required for the present, reduced pop-ulations of marine mammals in the North Atlantic is of the order of 50–60 million metric tons per year, compared to approximately 15–20 million metric tons taken by humans. This comparison is in part misleading, because much of what marine mammals consume are things we would not eat. Still some difficult questions emerge. Shall we provide (that is, not catch) enough fish to enable marine mammal populations to grow back to the level that may have pre-vailed before their hunting began? Or to some intermediate level? Beyond that level, shall we cull their populations again? This is im-portant: if we restore the ocean's ecosystems to what they were, say in 1950, some marine mammal populations are going to expand. We must expect that healthy ecosystems will produce healthy populations of marine mammals at the top of the trophic pyramids. If we are serious about marine ecosystem management, we must accept the keystone roles played by such predators in the maintenance of marine ecosystems, and this presents difficult choices.

It is clear that it will not be easy to restore the North Atlantic. Yet we have to restore its ecosystems if we wish to recreate an environment for sustainable fisheries.[8] Earlier, we mentioned some biological trends that must be turned around: the loss of large, long-lived fish; degradation of habitats;[9] and the "shifting baseline syndrome" that makes otherwise intolerable changes acceptable to successive generations. The reversal will take many years, if not decades or even centuries.

Moreover, ecosystem management involves more than just regulating the fisheries sector, particularly in coastal ecosystems, where all manner of enterprise, shipping, and water-use sectors intersect and compete for use of ecosystem resources. The intent of the concept of marine ecosystem management is to place ecosystem integrity or health as the primary consideration in all management decisions that affect the ecosystem.

From an economic viewpoint, the ecosystem restoration and management can be seen as an investment in a natural resource. As in any investment project, sacrifices have to be made today in order to enjoy benefits or payoffs in the future. The investment program to be undertaken, it should be noted, calls not just for investment in single natural resource assets such as a single population of fish, but rather for investment in a set, or portfolio, of natural resource assets, that is, in an ecosystem. Like the manager of a financial portfolio, one must be concerned not only with the expected return on the portfolio, but also with how risky the portfolio is. An ecosystem investment program could both substantially enhance the return on a natural resource asset portfolio and substantially reduce the risks associated with this portfolio.

It is not our purpose here to prescribe a comprehensive approach to managing marine ecosystems. That will depend on local conditions, including the state and type of ecosystem, its species composition, as well as a host of socioeconomic and political factors.

However, we can point out some solutions that are promising, though they have thus far been given inadequate attention in the vast majority of fisheries and ecosystems in the North Atlantic. These solutions fall into three broad categories of aims:

1. Reducing fishing effort.
2. Transforming the market for fish.
3. Transforming governance.

The tools and approaches presented herein, if used in combination with other, existing measures, offer the best, and perhaps only, hope of reversing the dramatic downward trend in North Atlantic marine life.

Reducing Fishing Effort

Overfishing is the main issue in North Atlantic fisheries. Addressing this issue vigorously, including the overcapacity of fleets and the subsidies that fuel it, will cause many related problems to fall away. Curbing overfishing "will cause some economic and social pain at first,"[10] but would reinforce the effects of other management measures and the political will to restore the North Atlantic ecosystems. If the benefits of reduced fishing effort are clear to fishers, managers, and other stakeholders, there should be better compliance and more public support for stewardship of the oceans. One simple solution to overfishing, of course, is to stop fishing entirely, which would let us step back and admire the majesty of the North Atlantic as a vast aquarium.[11] Our dilemma is that we want to have our fish and eat them, too.

Although it is inconceivable that fishing could stop altogether, there can be less fishing. In fact, first and foremost there must be far less fishing, fishing that should aim at catching only the right kinds of fish, only the right sizes of fish, and only a certain amount of each kind of fish, while avoiding nontarget fish, marine mammals, seabirds, turtles, and damage to the seabed.[12] Calls for these kinds of measures are already standard rhetoric in the conservation literature and the mass media.[13]

Fisheries managers have employed a host of regulatory mechanisms to reduce fishing pressure, including restrictions on the number of vessels, the number of licenses, the number of days at sea, and the gear types and mesh size fleets are allowed to deploy. These tools typically involve effort or capacity limits. Also known as input controls, these tools restrict what goes "in" to a fishery. While these tools achieve some desired effects (for example, gear restrictions help to reduce bycatch), they have not proven successful in reducing the capacity of the fleets to catch fish, and hence have not reduced fishing pressure. Yet there are tools we can use, and new approaches we can take to get the job done.

Quotas

The term "quota" refers to a certain quantity of fish that a fishery is allowed to catch, as determined by a regulatory agency, and belongs to the class of regulatory measures called "output control" because it refers to what comes "out" of a fishery. Quotas have very different impacts on fleet structure and operation, depending on how they are allocated. Thus, quotas expressed as total allowable catch (TAC) (without this catch being pre-allocated to the different, competing components of a fleet) usually have the effect of exacerbating competition between the fleet components, leading to a massive build-up of effort. This form of perverse incentive (see Chapter 3) can, and did in some cases culminate in "derby" fisheries, in which an entire annual quota was taken in less than a day of frantic fishing upon opening the season, by an enormous fleet largely idle for the rest of the year.

Other forms of quotas allocate a preset fraction of the TAC to different fleet segments, allowing these to take their part of the catch over an entire season or year, thus allowing massive fleet size reduction, and optimal fleet deployment. Such quotas may be allocated to either individuals, or to groups/communities. Different perceptions exist as to the legal nature of allocated quotas, which some consider enforceable property rights (just as the right to one's house), but which others consider a privilege, revocable as deemed appropriate by regulatory agencies. Some are regarded as inalienable, as is often the case with quotas allocated to communities.

Individual quotas are often held to be the general exception to the poor record of management measures, and ostensibly have the best potential to constrain the growth of fishing effort. Individual quotas have been in general use only since the late 1970s.[14] In the North Atlantic, Iceland has the most experience with their benefits and problems. In the U.S., four fisheries (Alaskan halibut, sablefish, wreckfish, and surf clams/ocean quahogs) had individual *transferable* quotas (ITQs, see below) before a moratorium on this specific form of individual quotas was declared in 1996. In the case of Alaskan halibut, the quota did resolve the previous issue of overcapacity alluded to above.

Nevertheless, catch quotas require good advice on the state of the fish populations in question, and accumulating evidence shows that our knowledge of the populations is faulty at best. In addition, quotas, like taxes on landed value, can create the incentive to under-report actual catches. Thus, for example, fishers will discard excess quota species (in amounts that are, of course, not reported) so they can continue targeting other kinds of fish after the quota on one has been reached. Especially worrisome is the practice of "high-grading" at sea. When boats are facing quota or landing restrictions, the introduction of onboard size-sorting machinery allows fishers to select only the favored sizes and discard massive numbers of less desired sizes. To suppress this, onboard sorting machinery has been banned by a number of countries, but is still legal in some of the major fishing nations around the North Atlantic.

Recently, there has been a move toward privatization in the form of individual transferable quotas (ITQs). The key term is "transferable," i.e., ITQs can be sold and bought. In theory, ITQs reduce fleet size by giving less profitable fishers the opportunity to cash out of the fishery by selling their quotas to more profitable operators. Competition to take the last fish before someone else does is reduced, hence providing the quota owners the opportunity to reduce their fishing effort as well. Thus, ITQs should lead to a more rational, stewardship[15] approach to the management of fisheries.

However, there are significant concerns about ITQs. Perhaps most importantly, they tend to encourage large-scale interests to dominate the industry. In the mid-1990s, two of the biggest holders of ITQs were the National Westminster Bank of New Jersey and the

American affiliate of KPMG, the largest accounting firm in the world.[16] This leads to disfranchisement of local, small-scale fishers, without whom sustainable fisheries are hard to conceive. We shall return to this theme.

In the U.S., the quota issue concerns the Public Trust doctrine, a common law doctrine (judicially developed rather than statutory), which includes the principle that the resources of the seas under U.S. jurisdiction belong to the public and that the government holds them in trust for the public. This perspective reinforces concerns about giving resources to private interests; confers a continuing duty on the government for responsible management of marine resources; strengthens the principles of the Magnuson-Stevens Act of 1976 that individual quotas are a privilege and do not create property rights; and implies that giving exclusive rights to a fishery should be accompanied by compensation to the public. It is these concerns which led, as part of the 1996 reauthorization of the Magnuson-Stevens Act (or "Sustainable Fisheries Act"), to a moratorium on new ITQ programs, which still stands.

One resolution to this may be to get away from the idea that quotas should be allocated in perpetuity in the first place, e.g., they could as well be auctioned off annually. Such auctions may well be regarded as a tax on fish, but unlike a flat fee imposed on fishers, the price is determined by their willingness to pay for the fish they catch. Indeed, this would not only resolve the right vs. privilege issue, but also resolve an equity problem that has also hindered the broad use of ITQs, i.e., the question of to whom they should be "given" in the first place.

A recent bill before the U.S. Congress seeks to reintroduce individual quotas on an individual fishery basis; the quotas would not be transferable, except in unusual circumstances. These nontransferable quotas, known as IFQs (for "Individual Fishing Quotas"), would be annually auctioned off, thus both generating revenues for management of the fishery and simultaneously restricting the number of boats in the fisheries thus restructured. Those who paid for the right to fish would cover the costs through savings in fishing capacity and effort.

A committee of the U.S. National Research Council recently found that it is not possible to set national standards with regard to quotas because of the vast differences between fisheries, and recommended that quotas should be evaluated in the context of individual fisheries at the discretion of regional councils, rather than imposed by nationwide legislation or regulation.[17]

Vessel buyback and destruction

Given the magnitude of the overcapacity in North Atlantic fisheries, vessel decommissioning must become a routine component of fisheries management, just like the decommissioning of weapons is a necessary component of any transition to peace (e.g., between the U.S. and the countries of the former USSR after the end of the Cold War, for nuclear-tipped missiles and tanks, or personal firearms after the end of the "troubles" in northern Ireland). Two sets of perverse incentives will have to be avoided in the process: (a) those leading to the redeployment of retired vessels into new fisheries, or other areas, and (b) those leading to the reinvestment of the money from buyback schemes into new fishing capacity (new vessels, gear, electronics, etc.).

Avoiding redeployment of retired vessels is straightforward in principle. It involves not only retiring a boat's license to fish, but physically destroying its hull and engine.

This can be achieved, as was done for ex-Soviet tanks, by pouring concrete into them. Alternatively, excess fishing vessels could be publicly sunk, i.e., turned into artificial reefs, encouraging SCUBA diving in shallow waters, and adding valuable habitat to deeper waters.

Various programs can be imagined, similarly, to avoid the money from buybacks being used for purchasing new vessels, or being used as collateral for loans toward such purchases. One of these would be for the money from buybacks to be applicable only toward one's retirement funds. Indeed, a generous program of this sort, in combination with auctioning the quota of the remaining vessels (see above) may help resolve, via an early retirement program, the painful transition issues that have so far blocked the debate on capacity reduction.

Thus, while such programs may be challenging to put in place, they can be designed to meet those challenges and can do so equitably. They are essential to attacking the overcapacity problem.

Marine reserves

Areas and/or seasons in which no fishing is allowed offer the means to protect fish during the most vulnerable stages of their life, times when they should not be hunted, especially the spawning and nursery stages. While seasonal closures of parts of some fishing grounds have been largely ineffective, many fisheries scientists are coming to the conclusion that *permanently* closed areas, sometimes called marine reserves (also known as "no-take" zones), are essential to ensure survival of heavily fished species. Marine reserves are an ecosystem tool par excellence,[18] rather than simply a way to reduce fishing pressure, and are discussed in detail.

Marine reserves are an attractive instrument for ecosystem restoration. The basic idea is to close to fishing a part of the present fishing grounds, usually the part that contains the spawning or nursery area, or even just a part of the fishing grounds where fish can pass through their life cycle unscathed and increase the numbers of larvae that might later provide increased recruitment of fish for nearby or even distant fisheries. With quota systems, gear restrictions, and other management mechanisms, the problem of enforcing restrictions is an Achilles heel. But one of the advantages of marine reserves is that they are in principle easy to police—by satellite, using tracking devices on fishing vessels. Tracking devices have been used for some time, particularly in remote fisheries. U.S. scallop boats working the Georges Bank, which include two areas closed to fishing, are required to carry these devices. Long-distance trawlers from the Faeroe Islands are also monitored in this way, and there is discussion on their future use in the European Union. Thus, marine reserves not only are the tool most beneficial to marine ecosystems, they are also eminently practical—if we can muster the political will to put them in place.

Fishery benefits

The benefits of marine reserves in terms of fisheries would vary enormously from fishery to fishery. Recent evidence[19] that for some species of tropical coral reef fish at least, a good percentage of individuals originate from spawnings on the same reef, suggests that a reserve on a reef mainly benefits that reef, with little impact on reefs further downstream. Nevertheless, such benefits to the home reef and so to the local fishers (rather than seeing closure of their reef benefiting mainly down-current reefs and fishers as was previously thought) are very real, as shown in a long-term Philippine study.[20]

In the North Atlantic, it seems that for cod, only a very large reserve, comprising 80% of the fishes' range, would have prevented their 1992 collapse in the Newfoundland/Labrador area. However, a much smaller reserve would have sufficed earlier, before the population was depleted.[21] Now, after the collapse, even with a 100% reserve (i.e., a closed fishery), the cod populations will probably take substantially more than 15 years to rebuild to even their poor levels before the collapse of the fishery.[22]

Farther south, in the U.S. northwest Atlantic, closed areas have become important elements of fishery management programs for regulated groundfish.[23] Two large areas of the Georges Bank totaling nearly 11,000 square kilometers and an adjacent area of about 6,000 square kilometers became protected from trawling and dredging at the end of 1994. Surveys since then have shown that the closures have been a major contributing factor in improved conservation and partial rehabilitation of depleted groundfish populations on the Georges Bank. Scallop biomass in and adjacent to the closed areas has increased 14-fold.

A de facto marine reserve was created around Cape Canaveral when the area was cordoned off for the U.S. space program. That area has been recently shown to be the only remaining source of trophy fish for the sport fishery.[24] But the numbers of marine reserves are small, and their size small. These cases may show just the tip of the iceberg of potential benefits that might be attained if we really followed the dictates of the available scientific research. In the

Caribbean, for example, it is estimated that between 20% and 40% of the area presently fished must be protected.[25] Even the Great Barrier Reef Marine National Park, Australia, often viewed as a textbook example of sound, sustainable management, is impacted by destructive fishing practices (who would believe that parts of the Great Barrier Reef are trawled for shrimp?), prompting a recent declaration increasing the proportion of no-fishing areas there.

One promising approach to increasing the area protected is to build a network of small marine reserves, rather than one or a few large ones.[26] Proponents, from the U.S. National Center for Ecological Analysis and Synthesis, argue that a network would create more variety of habitats in a given area and a greater perimeter per unit area. The latter is important because it is around the perimeters where fishers derive the most benefits of the reserves in terms of catches. But one problem with this approach is that large individual areas may be needed in order to maintain species richness or biodiversity, as on land. This is certainly the case in coral reef systems.[27] Another issue is the difficulty of controlling many small reserves versus a few large ones.

Perhaps the most important aspect of marine reserves is their ecosystem benefits. Undisturbed by fishing, marine ecosystems, particularly those parts impacted by bottom trawling, will grow back, although perhaps not to their original or pristine state. Nevertheless, lost bottom habitats will return and in time be re-colonized by former or new types of occupants. That much is sure, whether it is the restoration of a large tract of open ocean bottom or the re-growth of a reef.

The main point is that reserves will allow the biomass of the respective ecosystems to increase and, in so doing, raise up to their former size and shape those squashed trophic pyramids to which we likened the food web of a degraded marine ecosystem. An increase in trophic levels would be evidence of restoration as the ecosystem regained its integrity and health.

As with other aspects of conservation in this ocean, one's perspective is crucial, as we attempted to show with Figure 3. By 1995, there were 89 existing marine reserves and 12 proposed in the northwest Atlantic, and 41 existing and 12 proposed in the northeast

Atlantic.[28] However, less than 0.5% of the seas are included in reserves, and a lot less than that is actually protected from fishing: e.g., two thirds of the "marine protected areas" in Canada offer no legal protection.[29] Were we to turn around the debate about marine reserves and instead discuss, say, Allowable Fishing Areas, we might achieve a more level playing field in terms of defending marine ecosystems. As Figure 3 shows, the Allowable Fishing Areas in the North Atlantic are at present virtually 100%.[30] The potential of marine reserves in this ocean has thus hardly been explored.

Using the precautionary approach

Permanent marine reserves and the buying out of fishing capacity, whether through quota systems or vessel buyback and destruction, has the virtue that once these measures are in place, they become permanent features of the landscape—vessels are permanently removed from the landscape, and like national parks in the U.S., the reserves become treasured features. This makes them less susceptible to political and economic pressure, and means that they are not subject to the kinds of scientific uncertainty that attends stock assessments and the setting of TACs. But these tools alone cannot do the job. They must be used in conjunction with traditional tools. And the traditional tools must be applied with caution, allowing fisheries to grow incrementally toward sustainable yields, rather than aiming directly for maximum yields.

The potential harm that a given level of fishing on a fish population has on an ecosystem must always be the paramount consideration. Current regulatory approaches put the burden on the regulator, i.e., the government, or governments in an international instrument, to prove that fishing will have an unacceptable impact. Instead, industry should be required to furnish proof that fishing at a certain level does not harm the ecosystem. Such a reversal of the normal roles of government and the governed in the fishing industry could help simplify management by the government agencies concerned, thus allowing them to focus their research on issues of ecosystem restoration.

The fundamental features of a precautionary approach as it applies to fisheries are as follows:

- Erring on the side of caution in deciding allowable catches when the information is less reliable or uncertain.
- Acting to minimize the uncertainties of estimates of fish population sizes for management purposes and ensuring that management is robust in the face of remaining uncertainty.
- Giving priority to conserving the productive capacity of the resources where the impact of resource use is uncertain.
- Taking steps to protect critical habitats (such as spawning or nursery areas).
- Ensuring that the capacity of the catching and processing sectors is commensurate with sustainable levels of resource exploitation.
- Evaluating management options using the best methods and data available (especially prior to fishing or before restoration of populations) in consultation with all stakeholders (i.e., not only fishers, but also civil society).

An example of the use of the precautionary approach in the North Atlantic is that of the Norwegian spring herring fishery, which collapsed in 1970 to 0.1 million metric tons. Catching was allowed to resume in 1984 after a management plan was developed, and when the spawning stock biomass (SSB) reached 0.5 million metric tons. Only low-level fishing was allowed, even after the spawning stock biomass reached 2.5 million metric tons in 1994. The plan calls for only gradual increases in annual catch until the spawning stock biomass reaches its estimated maximum of 6 million metric tons.[31] We consider this a good plan because the stock is intended to rebuild to a high biomass, large enough for the stock to be able to sustain both fishing pressure and environmental fluctuations. This contrasts markedly with northern cod off eastern Canada, for example, for which catch quotas are still set (despite a "moratorium"), which is one reason why their stock is not recovering. Other reasons are that juvenile cod are caught as bycatch of other fisheries, and that many of these fisheries catch fish and invertebrates upon which cod normally feed.

The precautionary approach is not a new idea, nor is its application novel. But it is neither widely nor consistently enough applied. In the context of the economic demands upon fisheries, it is often difficult to argue for precaution, when people's livelihoods are at stake. That is why fisheries policy must also address the market pressures that drive overfishing.

Transforming the Market

The present market value of a commodity should be an indicator of the supply of that commodity. If supply is declining relative to demand, the price should rise and thus discourage consumption. But the present market value of fish is no guide to the status of fish populations in the ocean. The public has no direct warning system or incentive to seek alternatives to fish. This is in contrast to, for example, oil prices, which change as a reflection of potential scarcity (real or political) of that commodity, particularly in countries where oil is imported for the manufacture of automobile fuel, and where foreign debt, in which oil is a major factor, rises or falls overnight in relation to exchange rates.

However, like oil production and distribution, the fishing industry is also heavily subsidized. This means that prices are artificially low in both cases, and in neither case is the full social cost—which all of us pay, in terms of the costs to health, our air, our water, and our ecosystems—accounted for in the marketplace. Until we change this, continued demand will fuel continued destruction of the resource. There are several ways of reversing this situation.

Reducing subsidies

A careful look at existing subsidies shows that half of the 2.5 billion dollars spent in the North Atlantic produce negative results. Or put differently: over a billion dollars of taxpayers' money is spent by the countries bordering the North Atlantic on measures which contribute to making the long-term prospects of fisheries worse than they already are.

Of course, once in place, subsidies are notoriously difficult to remove, as they create parties vested in their perpetuation. A medium-term approach while dealing with these issues would be to advise the public of the extent to which the fish products they choose have been subsidized. This could be done by labeling the products accordingly, as is required in Europe of products that contain genetically modified organisms. This could both create political awareness of the issue and reduce market demand for heavily subsidized fish.

A related form of labeling, called "eco-labeling," will be revisited later in this book. It is widespread in environmental matters on land,[32] e.g., for forestry products. In 1997, there were eco-labeling schemes in 17 countries, and the European Union is developing a community-wide scheme for its member countries. The goals of such labels are to provide more information to consumers about the environmental effects of the product being consumed; to raise environmental standards; and, sometimes, to give producers in the issuing country a competitive advantage over other producers.

Although it can be argued that subsidies are not directly related to environmental issues, a scale of rating fisheries from say, A+ for completely unsubsidized and A where the costs related to fisheries management make up the subsidy, to F for heavily subsidized fisheries, would provide seafood consumers with an easily understood message. Any resulting changes in consumption behavior would be a strong signal to fisheries managers concerning the public's view about this issue.

Energy and/or carbon taxes

There is, as we have seen, a great range in the amount of fuel needed to catch a given amount of fish. Some fisheries appear relatively fuel-efficient while others are abysmal in this regard. Based on this kind of information, ratings could be placed on the products as to their energy efficiency, as we do for air conditioners and other machinery. We do not have the choice, of course, of targeting only those types of fish that we can catch with low expenditures of energy because we would soon run out of desirable fish. However, an eco-labeling scheme of this type could reach an equilibrium over time.

The first result of an energy efficiency label might be more effort being directed to such fisheries. This could cause increasing levels of energy intensity as the fish become scarce. A good monitoring program would pick up this increase and revise the energy efficiency label, perhaps directing in this way our choice to other fish. Catches of the original fish might fall as it lost popularity; its populations would then rise and the cost of catching it decrease again, resulting in a more favorable energy efficiency rating.

Alternatively, instead of an incentive for efficiency, imposition of penalties for energy inefficiency could be applied to fisheries. Carbon taxes are used in several European countries to begin offsetting global warming. For example, in Norway, the offshore oil industry is taxed US$38 (in 2000) per metric ton of carbon dioxide, the major greenhouse gas, released into the atmosphere through the burning of oil. Similarly, a tax on fuel consumption of fishing vessels, which, taken together, represent a significant level of carbon dioxide emission through fuel burned (for example, fishing vessels account for one third of all such emissions in Iceland), would represent not only a disincentive for energy-wasting fishing techniques, but also encouragement to reduce fuel use and thus greenhouse gas emission.[33]

There are great differences in the cost of fish protein in the different fisheries, a product of the protein content of the fish caught and the energy used. A protein efficiency rating on that basis (as distinct from a protein content description for nutrition labeling purposes) would be another or alternative guide of interest to the public, particularly with the rising interest in health matters and the increasing appearance of nutrition information on food products.

Educating consumers

Nongovernmental organizations are taking on watchdog roles and are themselves becoming quasi-institutions, e.g., Greenpeace[34] and the International Southern Oceans Longline Fisheries Information Clearing House (ISOFISH), mentioned earlier. These organizations use the "stick" approach by exposing bad practices. Greenpeace uses public campaigning to draw attention to bad fishing practices, while ISOFISH publishes information on offending vessel owners and companies.

The Marine Stewardship Council[35] uses a "carrot" approach called "eco-labeling." It seeks to reward fishing companies that follow good fishing practices by allowing them use of a logo or label to this effect on their products. Dolphin-friendly or dolphin-safe tuna labeling is one such logo, which has been accepted and understood by the public. Large tuna purse seiners were routinely killing large numbers of dolphins that became trapped in their nets; the vessels used dolphin sightings to locate schools of tuna that often travel beneath the dolphins. In the eastern Pacific Ocean, these vessels now have to employ special devices in the nets to release the dolphins.[36]

These approaches have their critics but have succeeded well in bringing the overriding issue, the decline in the quality of our fisheries and seas in general, to the attention of the public.[37] A parallel can be seen in the labeling of nutrition content in many processed and micronutrient-fortified foods. Governments had expressed concern about the impact, particularly on international trade, of fish product eco-labeling initiated by nongovernmental organizations. At the same time, fish certification schemes are being used by some regional fishery management bodies.[38] Whether or not government should take on this role, the public has the right to full knowledge of the seafood products they consume.

The Audubon Society's Seafood Lover's Almanac[39] that rates commonly eaten seafood in the U.S. according to the status of their fisheries, that is, the risk of collapse of the fished populations, is a good example of a less direct approach. Another is that by *National Geographic* magazine and the Ford Motor Company, who in January 2001, distributed to every school in the U.S. ten free copies of a world map showing nearly one thousand "ecoregions," with a focus on the "richest, rarest, and most at-risk areas."[40] As well, other groups in the U.S. recently brought the problem of swordfish overfishing to public and political attention through some 700 prominent chefs who took swordfish off their menus.[41]

Nonconsumptive uses of the oceans

A market-based process that supports conservation is the development, throughout the world, of the bird- and whale-watching industries. By 2000, the latter industry was worth at least US$1 billion, with some nine million people in 87 countries taking part. For the North Atlantic, the industry was worth an estimated $80 million in 1998 in direct expenses of whale-watching tourists; $343 million when indirect costs are added. With whale-watching growing at 10–20 percent per year, present values are undoubtedly considerably higher.[42] Although tourism is not without impacts, these pale by comparison to those of fishing, and moreover helps to build a political and economic constituency for healthy ocean ecosystems.

Accounting for future generations

Reducing fishing operations to the point at which profits are maximized—the maximum economic yield—is a more conservative measure than aiming, as do most guidelines, at maximum biological yield or keeping fish populations at levels that will provide that yield. As virtually every study of fish population biology and economics shows, the highest long-term profits are to be made by catching fewer fish than the maximum that the oceans can provide in the short term. Can such a reduction be adequate to protect the ecosystems? The answer is probably not. However, effort reductions toward the levels generating maximum economic yields would certainly be a step in the right direction, especially if we account for future generations by using an appropriate discount rate.

There are obvious gains to be made by moves to save or restore ecosystems, but they may not be obvious in an economic sense unless we consider our children and theirs, or put differently, unless we include intergenerational equity in the way we calculate net discounted benefits. For instance, even if we undergo the short-term social and economic pain of closing all or parts of some fisheries, in all likelihood the value of the restored fisheries, once reopened, will fade away after some years due to the way we currently discount flows of net benefits from those fisheries. However, if we discount

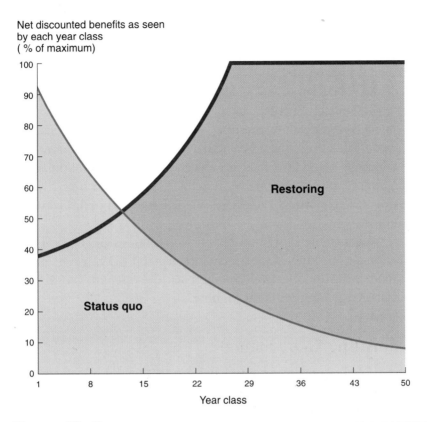

FIGURE 28. CLEAR ADVANTAGES FOR FUTURE GENERATIONS IN RESTORING MARINE ECOSYSTEMS.
Net present value of an ecosystem, as seen by each of a succession of 50 human-year classes, with benefits represented by the area under each curve.

The flow of net benefits as seen by each year-class is calculated using its respective discounting clocks, that is, the benefits to each year-class start being discounted when that year-class comes into existence, over its assumed 50-year life span; benefits are then summed over all classes, and expressed in percent of maximum possible, i.e., in relative terms. From Sumaila (2001).

the flows of costs and benefits from the perspective of all generations, both current and future, the picture changes.[43]

As an example, Figure 28 shows how successive generations, here represented by 50-year classes, value the benefits from restoration. It can be seen that, based on the net benefits each generation perceives, earlier generations would come to the conclusion that restoration is not economically sensible. But later generations would definitely find restoration to be a sensible proposition. When the interests of all generations are taken into account, restoration becomes the preferred course of action, since its benefits (shown as the area under the net present value "restoring" curve) are much larger than those under the status quo (area under the "status quo" curve). Figure 28 brings to the fore, in a very clear way, the important point that choosing to restore marine ecosystems is a choice to invest in the future, while choosing business as usual is a decision to disinvest in the future.

When weighing the costs and benefits of fisheries policy decisions, national governments must take into account the interests of future generations by effectively rolling back the discount rate for each successive generation as shown here.

Transforming Governance

As outlined in the previous chapter, a host of national management tools and institutions, as well as international treaties, conventions, and commissions have been developed to protect fish stocks. The kinds of tools and initiatives that need to be undertaken at the national level—implementation of marine reserves, the reduction of fishing capacity via buybacks, reduction of fleet sizes, and the improvement of individual quota systems—have been discussed above. But they are not enough. Distant-water fishing fleets pose a threat on both the high seas and within the exclusive economic zones (EEZs), particularly to straddling stocks. Moreover, cooperation among nations is needed to ensure that national standards are consistent and coordinated. Overfishing by one nation can harm adjacent ones.

However, the international instruments relating to fisheries in the North Atlantic have been found wanting in various ways. Indeed, there seems little connection between their stated objectives and the state of the fisheries that are covered by their mandate. Nevertheless, they are potentially useful in bringing a broad array of knowledge (from the countries involved) into mutual fishing concerns and in providing contracting countries platforms for negotiating settlements of disputes. Beyond that, their benefits are not obvious, in that although some regulations are designed to benefit all parties (where the sum of individual national contributions is enjoyed by all), some can reduce the benefits to all parties (where the level of compliance of the weakest link is the only level that can be enjoyed).[44] As pointed out earlier, however, even 100% compliance to a poor convention will not help to rescue a fishery from becoming depleted.

The FAO has developed a Code of Conduct for Responsible Fisheries (1995), a unique voluntary arrangement, in which any country or entity worldwide can adopt the principles contained in the code. It was developed to help countries deal with their rights and responsibilities in the legal regime following the declaration of exclusive economic zones (EEZs, in the late 1970s) and the UN Convention on the Law of the Sea (1982), particularly in view of effects on fisheries of the subsequent increase in fishing fleets to exploit EEZs and beyond. The intent of the code is made clear in its first general principle: *"States and users of living aquatic resources should conserve aquatic ecosystems. The right to fish carries with it the obligation to do so in a responsible manner so as to ensure effective conservation and management of the living aquatic resources."*

Several international plans of action have been developed within the framework of the code, adopted by FAO and published, for example, on conservation and management of sharks and management of fishing capacity.[45] An International Plan of Action to Prevent, Deter and Eliminate Illegal, Unreported and Unregulated Fishing, also developed according to the code, was recently approved.[46] The plans contain guidelines for assessment of the problems and procedures for developing national plans.

The FAO code has the potential to serve as a template for sensible rules, especially because it explicitly endorses the precautionary principle, and thus goes beyond most of the other international instruments examined in the previous chapter. Although recognition of the code is an important starting point for the development of international instruments, the substance of those instruments must nonetheless explicitly address several strong and urgent needs. They must:

- Deal effectively with the almost universal need to reduce fishing effort (as discussed above).
- Reconfigure their articles, using, wherever appropriate, the ecosystem instead of individual fish populations as their basis.
- Establish legal catches and take greater efforts (and perhaps inducements) to ensure compliance.
- Be transparent in their operation.

At root, the role of these international instruments is the establishment of legal catches and the prevention of illegal catches, which account for a large proportion of fish caught (see Chapter 3). Thus transparency, a critical component of the kind of modern governance we discussed earlier, is absolutely necessary. Transparency means that the processes by which catches are determined, sanctions imposed, and other decisions are made is open to interested parties, including nongovernmental organizations, who represent a broad public interest in fisheries. This transparency ensures that all the affected parties have had the opportunity to understand and affect the outcomes of management decisions, which in turn improves compliance. A key objective, therefore, is the creation of institutions through which the weight of public opinion can be brought to bear to ensure ethical, sustainable behavior by nations and individual fishing firms on the oceans, whether within national exclusive economic zones or on the high seas. An example of such an institution is ISOFISH, discussed below.

Dealing with illegal catches

Illegal catches can be exposed. The experience of a nongovernmental organization, the International Southern Oceans Longline Fisheries Information Clearing House or ISOFISH, bears reporting here. ISOFISH, along with some other organizations and national agencies, has been tracking illegal catches and sales of the Patagonian toothfish,[47] served worldwide in restaurants as "Chilean seabass." Illegal catches in various locations appear to have exceeded by a factor of some 500% the legal quota for this long-lived, vulnerable fish, set by the Commission for the Conservation of Antarctic Living Marine Resources.

ISOFISH has raised the profile of this illegal trade through its World Wide Web site,[48] in which it lists more than 90 offending vessels and their owners, many with detailed records of their illegal activities. A former Undersecretary of State for Fisheries in Chile was named in a recent ISOFISH release as the "pirate king" of these activities in that country. The Chilean Government subsequently took action to expose the trade but many offending vessels were re-flagged to countries making a business of deliberately neglecting their duties under international agreements (see below for examples of such countries).

Identifying illegal fishing activities and accounting for them in revised catch estimates are ways to create the transparency needed for management measures that will minimize the impact of fisheries on the marine ecosystems. ISOFISH, funded by the Australian fishing industry and the Australian Government, is a good example of what can be achieved, but it is not the only one. Norway has recently begun to blacklist vessels fishing outside quota arrangements in international waters on populations regulated by Norway, or fishing contrary to regional or subregional regulations. The names of the offending vessels are displayed publicly on the Internet.[49]

Despite these developments, the overall trend in this matter is toward less rather than more transparency. Although concern for underreporting of catches is widespread in national and international agencies concerned with fisheries, their response to the information is often at variance with expectations. Some examples:

- The Chilean vessels mentioned above that changed their country home port received unquestioning support from port and trade authorities in Uruguay, Mauritius, Mozambique, Namibia, and Réunion (a French overseas territory).
- While Australian authorities were quick to release for this study information on vessels arrested for illegal fishing, Canadian officials refused to release similar material.
- In the area covered by ICES in the northeast Atlantic, estimates of illegal fishing are regularly made by ICES working groups and adjustments made to assessments of some fish populations, but the basis of the adjustments is not made public; nor, importantly, does the official ICES database reflect the working groups' data,

does the official ICES database reflect the working groups' data, though this problem is now being addressed.[50]

The reasons for hiding such information, which by its nature, often involves vessels of "foreign" countries, include political embarrassment of allied nations as well as local political disincentives, such as linkages between government and industry that are so strong as to preclude effective control of the latter.

Whatever its rationale, the result is tantamount to fraud, on a scale that has decimated whole populations of fish and depleted North Atlantic ecosystems. It has increased the price of fish for consumers around the entire rim of the North Atlantic, while in many cases providing subsidies for (among others) the very defrauders. Governments are abetting this fraud.

FAO has pursued this issue in a series of working group meetings through which it has further investigated "illegal, unreported and unregulated" catches.[51] The reports point a finger at flag states of vessels, noting that "existing international instruments have not been effective due to a lack of political will, priority, capacity and resources to ratify or accede to and implement them." Further, the reports note that although existing International Maritime Organization conventions do not cover the majority of fishing vessels, states have considerable scope for introducing domestic legislation to deal with foreign fishing vessels. They also call for flag states to link fishing vessel registration with an authorization to fish, and to take all practicable steps to prevent "flag hopping."

One result of the FAO meetings is a voluntary international plan of action to deter "flags of convenience" by providing countries with "comprehensive, effective and transparent measures by which to act."[52]

Reducing the scale of fishing fleets

Since the development of modern fishing fleets following World War II, the marine resource pie has been cut into two unequal parts: the large-scale or industrial fishing fleets, capable of taking fish virtually anywhere, and the small-scale fishing fleets that are mainly limited to inshore areas. Throughout much of the twentieth century, at least until the mid-1970s, the heyday of North Atlantic fisheries, large-scale industrial fleets were seen as the embodiment of rational use of oceans, and small-scale fisheries as messy, inefficient leftovers from pre-industrial ages.[53] Now, in light of the various collapses brought about by industrial fishing in different parts of the world, and particularly in the North Atlantic, a reappraisal is clearly underway, with small-scale fisheries seen as nimble and efficient. Indeed, we argue that, given our present knowledge of the different modes of operation and socio-economic drivers of small- and large-scale fisheries, the fishing effort exerted by the former need not be reduced as much as that of the latter fisheries.

In general, the owner-operators of small-scale vessels, given their long-term dependence on local, coastal resources, should be more easily drawn into conservation-orientated co-management schemes than the shareholders of distant fishing companies. In other sectors, more localized, more community-sensitive bodies are leading to "modern governance,"[54] which gradually replaces unwieldy and relatively bureaucratic government departments. This decentralization has spread beyond even local government, and increasing responsibility for components of community life is being handled by non-governmental organizations (NGOs) or the private sector, particularly in developing countries where governments lack financial resources. This leads to various forms of co-management, in which control is shared between government and the community, and in which, typically, nongovernmental organizations act as the facilitators on the ground.

One example from the North Atlantic commercial fisheries is the Lofoten in northwestern Norway, where the fishery is divided into territories, each supervised by a superintendent. Regulatory committees are made up of inspectors elected by the fishers. The system is said to have more than 90 years of successful operation. In general, such systems need a strong element of trust, and the challenge is to find enough common ground in the needs and goals of fishers to form binding agreements. Conflicts, it is felt, can be overcome by using fishing rights in ways with which the local institutions agree.[55]

Thus, in a new approach toward resource sharing, a first step should be to review the implications of the relative efficiency of the two major classes of fisheries—small scale and large scale—and examine where gains can be made. Examples were given earlier showing the differences between the sectors in employment rates, use of the catch (for human consumption or animal feed, etc.) and its relative value, and how reallocating catch and effort between the two sectors could improve the economic efficiency of both. A shift in the allocation of fish toward smaller-scale fisheries will present significant opportunities for more effective forms of management.

Recommendations: Leaning on the Firewall Between Science and Advocacy.

There is, quite justifiably, a firewall between science and advocacy,[56] and scientists who are perceived as involving themselves in advocacy will lose their scientific credibility, both among their peers and eventually among the public at large.

This does not absolve us, however, from the job of communicating our results and their implications to as wide an audience as possible.[57] Indeed, if we do not attempt to communicate what our results imply, others, sometimes with agendas of their own, will do it for us, and the result may be confusion, rather than informed debate.

Thus, we present here a set of specific recommendations emanating from the material presented in this book. They represent a set of solutions, but are not the only ones. Another set may possibly, and probably will be, derived from this material. With these recommendations, we are leaning on the firewall between science and advocacy. But, we will not break through, or attempt to get around that firewall. We do not, for example, intend for our recommendations to take the place of informed political discourse. Nor do we intend to inject our own values into the debate. Rather, our recommendations are intended to give a sense of the steps we must take if we take seriously the idea of conserving the oceans as a living, productive resource. The only thing our suggestions have to recommend themselves is that they are based on the findings presented in the preceding text, i.e., what we know about the state of the North Atlantic fisheries and the effectiveness or ineffectiveness of the measures taken to preserve them. There is nothing new about these five recommendations. What we are making, we believe, is a coherent case that they are plausible. *Something* must be done.

Recommendation 1
Fishing pressure on North Atlantic fish population and ecosystems must be drastically reduced, by a factor of three or four in most areas.

The massive reduction of fishing effort and capacity called for in recommendation #1 is the price we have to pay to avoid the single-species collapses illustrated in Figure 9, and the reduction in biomass shown in Figures 10 and 14. If we do not do it on our own, nature will do it for us, through further stock collapses.

Tools exist that can be used to achieve such effort reductions. Property rights via individual fishing quotas will do the job in some limited cases, as will unannounced decommissioning (buyback) programs, with subsequent destruction of the vessels in question. These programs will work only if they are implemented such that the political representatives of affected regions are prevented from using calls from their constituents as a pretext for not acting decisively. One example of legislation of the type required here is that passed to enable the closure of military bases in the U.S., where members of Congress could vote for or against the package as a whole, but not pick and choose the individual bases to be retained, which was an administrative decision. One very effective tool, in this context, is the abolition of subsidies. This is an issue that resonates across a broad political spectrum, and which should be straightforward to communicate to a wide range of audiences. There is no reason to believe that the public at large wishes to subsidize the destruction, by a relatively small industry, of resources that belong to all.

Care must be taken, in this context, to avoid exempting fishing fleets from the carbon taxes or equivalent instruments that will gradually begin to be implemented in the coming years, as we begin to introduce measures against the release of greenhouse gases into the atmosphere. Such measures will be undertaken internationally (whatever we might presently feel about the Kyoto treaty), and it is useful to think about fishing fleets in this context. Indeed, given the increased consumption of fossil fuel per unit-catch documented here, and the marginal profitability of many large industrial fleets, it may turn out that this issue becomes their Achilles' heel.

Recommendation 2
Large marine reserves, amounting to at least 20% of the ocean by the year 2020, must be established, pending the long-term transition toward a regime where specific areas are explicitly open for fishing, while the rest is closed by default.

"Sustaining" the fisheries that exploit the collapsed populations in Figure 9 would be foolish. These populations must be rebuilt, and the ecosystems in which they are embedded restored. Marine reserves are required for this, because without a place where they can maintain their original, diversified population structure and abundance, the long-lived fish that we exploit will not only fail to rebuild their previous biomass, but rather go extinct. Extinction is gradual. In the North Atlantic, many species of commercial fish species have already lost most of their populations. When the last population goes, a species is extinct. Avoiding this fate for the various species that are endangered will involve allowing not only their biomass to rebuild, but also the ecosystems in which they are embedded to regain their previous structure. Marine reserves, by rebuilding our capital in the bank, reduce the risk of massive losses when global climate changes start modifying the physical features of North Atlantic water masses.

Moreover, marine reserves will take us away from the edge of the cliff, and shrink a bit the bloated balloon that we previously used as metaphors for our present style of fisheries management. We therefore endorse the widely accepted "20/20" formula to place 20% of the ocean into marine reserves by the year 2020, though we fear that these arbitrary numbers will, in most parts of the world ocean, prove too little, too late.

The marine reserves should be large enough and located such

that they allow the rebuilding of marine mammals and seabird populations. However, in areas left open for fishing, explicit consideration will have to be given to the food required to maintain their recovered populations. The scientific tools exist to perform the required accounting for biomass. The major impediment here is the continued perception that populations of small fish that are not fished are therefore "underutilized," and should thus become the target of further fisheries expansion and "diversification."

Recommendation 3
Eco-labeling and other market-based efforts to move the fishing industry toward sustainable practices must be intensified.

For most of us, buying fish is the major way we impact the ocean (besides paying taxes that are used to build fishing boats nobody needs). In addition to the criteria used by the Marine Stewardship Council mentioned earlier, new ways should be found of making the public aware of the true costs to society of their choices in fish consumption, e.g., through an energy and/or protein efficiency label, displaying in markets and supermarkets the relative energy costs of catching different kinds of fish. Another carrot would be a subsidy rating on fish products to advise the public of how much they are indirectly paying for a product in addition to the stated price.

Recommendation 4
An effective regime must be designed and implemented to publicly expose deliberately unsustainable and illegal practices, and their perpetrators.

The "stick" approach should involve increased information to the public of illegal fishing activities, including outing of the perpetrators. Notably, the technique law-breakers use of reflagging their operation to countries that tolerate and thus encourage their activities could be turned against them, by posting details of their activities on web sites in countries where freedom of expression exists.

Criminal behavior by vessel owners and their crews, and negligence by governments and international institutions tasked with conserving and protecting public resources (not to speak of ecosys-

ecosystems), will have to be publicized as well. We cannot go on ignoring the greed-motivated destruction of public property.

Recommendation 5
Access and property rights in fisheries should favor smaller-scale, place-based operations, operating passive gear to the extent possible, and the fisheries should be run through co-management arrangements.

We do not recommend that small-scale fisheries should be favored over large-scale operations because of romantic notions of rugged small operators battling both the elements and anonymous corporations. Rather, we make this recommendation from the scientific evidence available to confirm the common-sense inference that local fishers, if given privileged access, will tend to avoid trashing their local stocks, while foreign fishers do not have such motivation. We are aware of exceptions and could provide caveats, but the principle holds. Moreover, small-scale fishers, both because of their geographic closeness to the resource they exploit, and because of their tendency to use passive gear (e.g., traps), tend to use less fuel and do less damage to marine habitats than their large-scale counterparts.

Co-management schemes are required, in this context, because only the fishers themselves and the communities in which they are embedded can ensure that everyone plays by the rules. Top-down control by central governments, the mechanism so far relied upon to ensure compliance, only leads to defiance, and a game of catch-me-if-you-can in which the resources and the ecosystems suffer.

Ultimately, these general recommendations must be applied with vigor and in detail by individual nations, international institutions, the fishing industry, and nongovernmental organizations. No one of the above recommendations is sufficient by itself, and no one sector can accomplish the task by itself. While one may quibble with one or another of our findings or propose alternative measures, one thing is abundantly clear. The North Atlantic Ocean is severely degraded and remains under attack.

We end our plea here and rest our case.[58] The ocean waits for our action, good, bad, or indifferent.

Notes

PREFACE

1. Rachel Carson's *Silent Spring* was crucial in helping people realize the ecological impact of DDT and other pesticides, even if it was their impact on *human* health that led to the regulation or banning of these compounds. Her other well-known book, *The Sea Around Us*, also conveys a strong sense of how tightly connected marine ecosystems are. However, she identified pollution (especially radioactivity) as the major threat, not fisheries. Here, jointly with Jackson et al. (2001), we rectify this.

INTRODUCTION

1. The notion of new mental maps is presented and explored in Pauly and Pitcher (2000), who present the historical, conceptual, and methodological underpinning of the *Sea Around Us Project* (see www.saup.fisheries.ubc.ca/).

2. An attempt to put values on the planet's ecosystem services (Independent World Commission on the Oceans [1998] concluded that marine ecosystems accounted for two thirds of the total global value, the terrestrial ecosystems the remaining one third (a fact that might seem obvious because the land comprises roughly a third of the total; yet we derive nearly all our food from the terrestrial third). This feature alone highlights the great importance of maintaining the integrity of the oceans' ecosystems. The annual value in 1994 of all the oceans' services was estimated at $21,000 billion. The Atlantic Ocean makes up 29 percent of the total ocean area and the North Atlantic somewhat less than half of that. Thus, we could say the North Atlantic maintains our environmental equilibrium to the tune of $3,000 billion per year. [Actually, the figures cited in the Independent

World Commission on the Oceans (1998) are: value of all marine ecosystem services—36,302 million hectares (ha) at US$577/ha/year = $20,949 billion/year, of which the open oceans, 33 million ha at $252/ha/year, contribute $8,381 billion/year, and the coastal seas, 33,200 million ha at $4,052/ha/year, contribute $12,568 billion/year. The total terrestrial services, 15,323 ha at $804/ha/year, contribute $12,319 billion/year]. Yet, these are vital services, without which we would all perish. Really, they are priceless.

Further, as well as what we call services, we have to remember the basic functions of marine ecosystems: that they support populations of fish, other animals (including birds) and plants; and at a higher level they support the structure of the communities of these animals and plants, and their diversity and that of their genetic resources (Ehrlich 1988). Again, a list of ecosystem functions beyond price. Beyond price, meaning they *cannot* be swapped for any amount of currency or alternative goods. Moreover, there is a huge psychological value of the sea that nobody has attempted to convert into currency. The oceans, particularly the nearshore seas that we most readily see, have immeasurable value as "places," reference points in the geography of our minds as much as of our eyes, moving one observer to point out regarding coastal seas that: "A sense of community encompasses not only one's families and neighbors, but one's place in the local ecosystem as well." [from a contribution in Anthea Brooks and VanDeever (1997)] This attitude takes us back to the point made earlier: that we are ourselves part of marine ecosystems by virtue of our fish-eating habits.

3. The strong interconnections between degradation of the environmental resource base and poverty, population growth, undernourishment, and erosion of civil society are well documented. Nevertheless, conventional indicators of living standards relate to commodity production and ignore the resources on which production ultimately depends. As suggested by Anita Kelles-Viitanen in an Asian Development Bank seminar (ADB 2001), there are clear needs to temper the optimism of such indicators with those relating to the state of these resources. To date, champions for the inclusion of environmental indicators in measures of the quality of life have explicitly or implicitly referred only to terrestrial issues or those related to the atmosphere (Ward and Dubos 1972; Henderson 1991).

The most widely used indicators of quality of life at present are those contained in the Human Development Index: longevity (life expectancy), educational attainment, and standard of living (real GDP per capita). The only environmental indicator in general use for this purpose is the annual rate of deforestation (see Appendix 1 of Bloom et al. 2001). Therein, the following are viewed as components of quality of life: nutrition, health, education, income, gender equality, fertility, political and civic and economic freedom, environmental quality, access to infrastructure, and access to information.

The OECD has one environmental indicator for marine resources: quantity of catch. By this indicator, countries with a large fish catch have a poor rating. We need something more realistic than that.

For the oceans, the North Atlantic in particular, the decline in mean trophic levels documented in Chapter 3 of this book is one such indicator. Like body temperature in humans it does not register all the problems of the fisheries, but its inclusion in assessing living standards or quality of life of human populations around the North Atlantic Rim would be a great step forward in bringing some sobriety to proponents of the current measures of economic "growth." That single value would mark our progress in either continuing the decline or fostering the rehabilitation of the ocean's ecosystems, ecosystems of which we are a part and on which we all ultimately depend; a single value that, over the years, would inform us whether we predators are part of the problem or of the solution with regard to restoring our ecosystems. This would help us address the intriguing possibility that the quality of life in the oceans is associated with the quality of our lives ashore.

A return to bountiful ecosystems in the North Atlantic would not necessarily imply improvement in civil society or in poverty reduction, but one can easily imagine that a societal attitude that fosters the health of the seas may also have spillover effects on other aspects of our lives. As pointed out by Pr. Karina Constantino-David at the Asian Mayors' Forum in Shanghai, June 2000 (ADBI and ADB 2001), the opposite extreme prevails in Asia, home to the majority of the world's poor and where the present form of "sustainable development" has been called a parasitic one, that "sacrifices the poor and the environment at the altar of the market and its promises of economic growth." And there may well be more than just quality of life involved. The present rates of species extinction and climate change could have serious implications for the survival of our own species. The oceans are major factors in climate maintenance and further marine ecosystem disruption can only exacerbate these trends.

Even if we manage to compensate in time for foreseen and unforeseen effects of further climate change, we still must deal with—redress—the long-term effects of marine ecosystem degradation due to fisheries in the North Atlantic Ocean. Outstripping the capacity of natural resources to support humankind is undoubtedly a major factor in the demise of previous civilizations (Wilson 1998). Present civilization is patently outstripping the capacity of the North Atlantic, which is one very large piece of the totality of the Earth's natural resources.

That ancient, not-so-naive question by Omar Khyaam "I often wonder what the vintners buy, one half so precious as the goods they sell" is relevant even here (the quote is from an undated edition of the Fitzgerald translation of *The Rubaiyat*, published by G.G. Harrap and Co. Ltd., London). As the continuing mismanagement of North Atlantic fisheries implies, governments today appear to be selling our own futures. Yet, as a general principle, we hold the oceans in trust for future generations. In a different sense of the word, we must establish institutions that are seen by both society at large and fishers in particular as trustworthy to take on the roles of defenders and protectors of the oceans. Society has to trust that industry, both large and small scale, will honor its obligation to conserve the oceans' resources; while fishers and the fishing industry have to trust that society (through properly enforced regulations) will protect rationally managed fisheries.

As shown in Chapter 3 of this book, there is an urgent need for transparency in fisheries—in vessel registration and operations, and in analysis of catch and trade data. Indeed, scientific credibility forms a key part of this trust matrix and here the record, in terms of collapsing fisheries (whether or not contrary to scientific advice) is not a good one. We must stop pretending that fisheries can continue to be fine-tuned to deliver, and retreat to more certain ground in order to regain the trust, if not the agreement, of the fishing community and the public at large. This requires admitting that the Ocean's ecosystems are in a state of decline. Addressing that broader issue in a transparent way is surely in accord with everyone's wishes as a means of maintaining or improving our quality of life.

CHAPTER 1

1. Quoted in Mowat (1984, p. 169).

2. The description of the evolution of the Atlantic Ocean is drawn from Charton and Tietjen (1989) and the Independent World Commission on the Oceans (1998).

3. The following information on environmental processes is summarized from de Young et al. (1999).

4. North Atlantic atmospheric circulation is impacted by a low-pressure center near Iceland and a high-pressure area near the Azores. The changes in the difference between these centers (the North Atlantic Oscillation or NAO) affect both air and sea temperatures. In the early 1990s, the sea surface temperature was colder than normal in the Northwest Atlantic and resulted in warmer than usual conditions in the Northeast Atlantic. There is a need to improve our knowledge of the NAO for fisheries management purposes (NRC 1999b).

5. Dyson (1977). The current that brings Gulf Stream water to Europe is the North Atlantic Drift, part of an inter-ocean "conveyor belt" that flows north in the North Atlantic until reaching the Arctic where its waters cool, sink and flow back south near the sea bed. This deep current splits into Indian and Pacific Ocean segments, rising to the surface in those oceans in a loop that brings it back along much the same route into the North Atlantic on the surface. In recent years there have been increasing amounts of freshwater entering the Arctic Ocean as a result of gradual melting of the polar ice, possibly due to global warming. This freshwater mixes with the North Atlantic Drift and makes the latter more buoyant. It has been theorized for some time—and in 2001, new evidence appeared that supports this—that a point may be reached where the mixed water could stop sinking. The conveyor would then cease to flow and the climate of northwestern Europe would cool by an estimated 5 degrees Celsius. The net effect could actually be an increase in ice cover in the North Atlantic (see: *New Scientist*, www.newscientist.com, 19 February 2001). But, after perhaps a few decades, the same cooling effect would reverse the melting of polar ice, decrease the freshwater input and turn the conveyor on again. The effects on fish populations of such temporary changes in current patterns are

presently unknown, although it seems likely that most would be affected negatively by the disruption of the physical mechanisms leading to the "triads" of local enrichment, concentration and retention that enabled large fish populations to maintain themselves, and which depend on the match between currents and coastal features they have evolved to exploit (Bakun 1996).

6. The description of biomes and biogeochemical provinces (BGCP) is from Longhurst (1995, 1998). Sherman and Duda (1999) describe a system of 50 large marine ecosystems (LMEs), recently expanded to 64 LMEs (http://www.edc.uri.edu/lme/default.htm). The conceptual basis for the integration of the BGCP and LME classification was presented in Pauly et al. (2000), who also provide more details on and a rationale for the mapping procedures used in this study.

7. More details are given in Christensen (1996).

8. Recent estimates of primary productivity for the entire ocean may be found in Longhurst et al. (1995).

9. Planktonic organisms such as krill (euphausiaceans) are increasingly being targeted, e.g., off Norway, as food for farmed fish, as are jellyfish, which supply the East Asian market, where they are considered a delicacy.

10. One good reason why we would want to get proper baselines of the marine life we have now, or once had, is because it is only when it is based on well-established baselines that the concept of sustainability, otherwise nothing but a feel-good concept, can be made meaningful. Indeed, unless firmly rooted in scientific, quantified knowledge of what we now have, or had, we will experience what has been called the "shifting baseline syndrome" (Pauly 1995). Herein, successive generations of naturalists, ecologists, or even nature lovers use the state of the environment at the beginning of their conscious interactions with it as "the" reference point, which then shifts as successive generations degrade that same environment (the story of the frog kept in water that is heated very slowly comes to mind here, and if we are not careful, we are going to get boiled as does the frog, by a runaway greenhouse effect).

11. Pitcher (2001a).

12. Mowat (1984) is an indispensable source for any discussion of past abundances in the Northwest Atlantic. The lobster weights cited below were also taken from this source (p. 203).

13. Mowat (1984); note that the slaughter of seabirds continues, now in the form of bycatch (Brothers et al. 1999; FAO 1999a).

14. Based on account of Palissy's life and work by J.M. Barrande in Tort (1996).

15. This section is drawn mainly from Dyson (1977).

16. The Domesday Book (survey of English lands and holdings ordered by William the Conqueror in about 1086) listed 76 tidal fisheries controlled by the crown. By 1154, when Henry II ascended to the throne, fishing rights had been granted on nearly every stretch of water to individuals, estates and town corporations. The Magna Carta ended the crown's privilege in 1215. Then, rights were divided among types of fish and shellfish; legal terminology kept lawyers arguing for

centuries. The House of Lords finally ruled that shellfish could be owned but swimming fish were a common resource and did not belong to anyone until captured. As early as 1320, the capture of immature fish was illegal in the River Thames (Dyson 1977).

17. From Coull (1993). North Sea herring were a staple food of both rich and poor in England and Europe from at least the Middle Ages (when the Hanseatic League founded its fortune on herring) until the mid-1800s. The archers of Agincourt went into battle on breakfasts of salted herrings (Dyson 1977).

18. English fisheries, in general, had greatly benefited from the introduction of ice and the expansion of British railways by the mid-1800s. Over a few years, fishing was transformed from a badly organized cottage industry to an immense capital-intensive operation involving steamships, railways and chains of food stores.

19. The first base for Norwegian whaling was in Finnmark, northern Norway. The failure of the traditional fisheries occurred at the same time as whale numbers within reach of the station there were becoming overexploited. In 1880, Norway banned the whalers from operating within a mile of the coast between January and May. The fishers were not satisfied with only a partial ban and continued agitation for a total ban. (Dyson 1977).

20. Information in this section is adapted from Dingsør (2001).

21. The herring collapses are mentioned in Myers et al. (1996b) as having the same causes—overfishing due to a high fishing mortality rate, and high discard rate of juveniles—as for the Atlantic cod collapse in 1992.

22. www.intrafish.com.

23. The 1990 report on EU fish populations was called the "Gulland Report"; the more recent report on population status is by the Commission of the European Communities (2000).

24. As reported in *New Scientist,* 27 January 2001.

25. Following current usage, we use the term "Inuit" for the people previously called Eskimos.

26. For all groups, apart from prairie communities, "the fish-hook and the fish-spear, the net, trap, and weir, were as indispensable at certain seasons as the bow and arrow." Eastern Canadian First Nations jigged from canoes in bays and gulfs, catching mainly cod, halibut, and salmon. In the north, canoes were used for trolling; and jigging was through holes in the ice of lakes to catch whitefish, trout, and salmon trout, and occasionally to catch rock cod through cracks in the sea ice during autumn and spring. Most groups also captured large numbers of sturgeon, salmon, and eels with torches and spears at night. However, most fish were caught using seine nets, even under the ice, as well as by bag nets and dip nets. The Inuits constructed stone weirs while groups farther south made weirs of timber and brush. Shell heaps or kitchen middens testify to the early consumption of significant quantities of molluscan shellfish (Jenness 1972).

27. Details in Kurlansky (1997, p. 17–29).

28. There was a period of more than 300 years of struggle for possession of the cod fishery, marked by "attacks and counterattacks, destruction and counter destruction, pitched battles, piracy, murder, and looting, war on both land and sea" (Johnstone 1977). Rivalry between English and French fishing settlements resulted in the destruction of all the English settlements by the French over the period 1696 to 1708. During this period, the British Parliament banned settlement other than in support of the cod fishery. Treaties from 1713 accommodated French fishers, although British sovereignty was not recognized until the Treaty of 1783 at the end of the American Revolution.

29. The term "cod" is applied to many species of gadid fish in the North Atlantic. The major one is the Atlantic cod (*Gadus morhua*). Arctic cod (*Boreogadus saida*) and Greenland cod (*Gadus ogac*) are minor species in the North Atlantic.

30. Groundfish is the term applied to the mix of bottom-living fish species caught initially by hook-and-line and traps, and later also by trawl.

31. The Newfoundland cod fishery was England's prime overseas fishery until Arctic trawling began in the 1890s.

32. Information in this paragraph is from Cadigan (1999).

33. Information on Canadian cod populations is from de Young et al. (1999).

34. Gulf of Maine fisheries development is taken from Alder et al. (2000); the halibut story is from Dobbs (2000).

35. No other explanation—and there were several—for the collapse of the cod population fits all the results of the tagging data (Myers et al. 1996a and 1996b).

36. Mowat (1984, p. 172), citing a pilot in the Royal Canadian Air Force, who flew patrols over Canada's coastal waters in the 1950s.

37. Myers et al. (1996b).

38. Hutchings (2000).

39. As witnessed and reported by Dobbs (2000).

40. Too much fishing, or "overfishing," is defined by the (U.S.) National Research Council (NRC 1999b) as "fishing at an intensity great enough to reduce fish populations below the size at which they would provide the maximum long-term sustainable yield or great enough to prevent their returning to that size."

41. As a complement to this history of devastation, especially of Atlantic cod, archeological evidence shows that the average size of cod landed from the Gulf of Maine decreased from one meter to one-third of a meter in the years following the onset of mechanized fishing in the 1920s. It is noteworthy that this came about after at least 5,000 years of stability of both cod and kelp forests, despite intensive aboriginal and European hook-and-line fishing. However, after the 1920s, cod declined in both numbers and size and their prey, sea urchins, increased and severely reduced the kelp forests until the urchins in turn were fished down (Jackson et al. 2001). In the Gulf of Maine, they are now fishing sea cucumbers. You cannot go lower than that in terms of fishing down marine food webs: sea cucumbers eat dirt.

42. By 1376—only some six years after introduction of the trawl—trap fishers petitioned the king [Edward III] about the decline of fish the trawl was causing by "destroying the flowers of the land beneath the water, and also the spat of oysters, mussels and other fish upon which the great fish are accustomed to be fed and nourished." And "by which instrument in many places the fishers take such quantity of small fish that they know not what to do with them; and they feed and fat their pigs with them, to the great damage of the Commons of the Realm and the destruction of the fisheries." (Dyson 1977).

43. Watling and Norse (1998).

44. See, e.g. Hall (1999).

45. That global fisheries catches have been declining since the late 1980s is a recent discovery (Watson and Pauly 2001, Watson et al. 2001a, b), not anticipated in global statistics based on official statistics (FAO 2000a). Detailed information on catch trends for each of the world's major ocean areas may be found in the *Sea Around Us* Project catch database (see http://saup.fisheries.ubc.ca/).

46. See Pauly et al. (2002) for more on this and related features explaining why they tend to be unsustainable, even when regulated.

47. This account of the role of bivalves is inspired from Jackson et al. (2001).

48. Jackson et al. (2001).

CHAPTER 2

1. See Giberne (1910), a popular account of life in the sea.

2. The key models of fisheries science, which relate fishing effort (or fishing mortality) to catch on a species-by-species basis, have a central assumption that is actually never met in nature: that we can isolate one species in the ocean from its predators and prey. Nevertheless, empirical evidence has shown the system of equations derived from these models to provide useful predictions in the field (Cochrane 2000). All interactions between fishers and fish, however, are of a multi-species nature. The effects resonate throughout the food webs within which the target species are embedded. The degree to which these secondary effects distort our perception of the interaction between fisher and fish is one major cause for the inaccuracy and uncertainty inherent in the results of these models.

3. McGlade (1999).

4. The software and its applications are documented in Christensen and Walters (2000), Pauly et al. (2000), and Walters et al. (1997, 1998).

5. See Christensen et al. (2001) for details on this, and Christensen et al. (2003) for the peer-reviewed version.

6. Information in this section is mainly from Heincke (1913) and Zeller and Pauly (2000, 2001).

7. The principal source was *FishBase 2000* (Froese and Pauly 2000; see also www.fishbase.org).

8. Previously, information on fish landings for the North Atlantic was only available (from FAO) on a very coarse scale—only four areas comprise the FAO divisions of the North Atlantic as defined here, with no further information as to where the landed fish were actually caught (see top panel of Figure 2). Ten percent of the fish landed remain unidentified, and an unknown, but probably substantial, fraction of the catch remains unreported because it was caught illegally, discarded, or landed by small-scale fisheries not covered by the statistical reporting systems of the FAO member countries bordering the North Atlantic (Pitcher and Watson 2000). The International Council for the Exploration of the Sea (ICES, covering the Northeastern Atlantic) and the Northwest Atlantic Fisheries Organization (NAFO) report catches, by "stock," in their own finer statistical divisions, but also fail to include all catches. Thus our emphasis on reconstruction of independent catch series (see, e.g., Coelho and Stobberup 2001).

9. The new "mental map" that will hopefully result from this is discussed in Pauly and Pitcher (2000).

10. The concept of the "shifting baseline syndrome" was introduced above. Here is an example of what drives it: "The North Sea which today's fisheries scientists grew up with, was very different from the North Sea before the trawl. The giant cod, halibut, and turbot we used to land are gone, replaced by smaller flatfish and scavenging dogfish and rays. The large sharks are an even more distant memory." (Roberts 1999). To take another example, a Canadian fisheries expert (John Hart, cited by Johnstone 1977, p. 22) stated that "by 1910 people engaged in the fisheries of the North Atlantic coast had no clear recollection of former conditions of excessive abundance or of sharp declines in the reward for fishing effort. Supplies of older fish had already declined. Acceptance of the situation, of operating in a stabilized fishery, was accordingly current in the early years of the 20th century."

11. National fisheries catch statistics in 1900, though not assembled and standardized by international organizations, were not necessarily less reliable than today. Indeed, they may have been more reliable in many cases, as the "perverse incentives" (NRC 1999b) now existing for fishers to misreport their catches (quotas, prohibitions on landing certain species or size groups, etc.) did not exist then, and dockside sampling was often done on a very systematic basis.

   The *Sea Around Us* web page (http://saup.fisheries.ubc.ca/) and Christensen et al. (2002) provide details on the construction of this map.

12. Actually, it is not really necessary, at least at first, to collate national catch statistics (usually published by the fisheries or agriculture ministry of various countries, and by the Department of Commerce in the U.S.) to estimate catches taken from larger areas of the Atlantic or other oceans. Various international organizations, notably the Food and Agriculture Organization of the United Nations (FAO), based in Rome, or the International Council for the Exploration of the Sea (ICES), based in Copenhagen, are tasked with collating, standardizing, and disseminating such data. As we will show, however, the data thus gathered often have serious drawbacks, and must be complemented by specific studies, if they are to adequately reflect reality. In this book, we indicate discrepancies

wherever appropriate, while remaining aware of the crucial work performed by the technical staff of these, and similar institutions.

13. The "catch maps" are constructed using a novel rule-based method described in Watson and Pauly (2001) and Watson et al. (2001a), which allocates catches by species to the 1/2 lat./long. degree spatial cells from which they can originate, given constraints provided by the distribution or range of the species and access rights by the reporting countries to the exclusive economic zones (EEZ) of countries where the species in question occur.

14. From Alverson et al. (1994).

15. Details in Rejwan et al. (2001).

16. A methodology for such work is given in Watson et al. (2000) and Pitcher and Watson (2000). Some 75% of Spain's swordfish catch in 1997 was probably illegal; catches of cod and whiting from western Scotland populations since 1991 appear to have been underreported by a factor of 30% to 60%; massive quantities of undersized cod are said to be landed in France; midwater trawlers in western Ireland probably exceed the quotas for herring and mackerel by 100% to 300%; 50% of the catch of Scottish purse seiners is said to be illegal; Humberside (UK) fisheries take about twice the reported catch; and a Spanish trawler, arrested in Canada, had a secret hold for unreported and illegal fish, which in total were the same quantity as the legal catch. The Peruvian anchovy fishery, the largest in the world up to the early 1970s, collapsed spectacularly in 1972. In the years preceding the collapse, it became clear that the official catch figures were massively underestimating the true catch. The official catch for 1970 of 12 million metric tons was at least 4 million metric tons less than the real catch (Castillo and Mendo 1987).

17. There are many gear types: if the truth be known there are probably nearly as many gear types and variations as there are fishers. But, they were grouped into a reasonably small number for this assessment: bottom trawls, midwater trawls; mobile seines, surrounding nets, gillnets and entangling nets, hooks and lines, trap and lift nets, dredges, grappling/wounding gear and harpoons and spears, and "other" gear.

18. This is based on the definition: *catch = fishing mortality * biomass.* Inverting this gives $F = catch/biomass$, with F renamed fishing intensity, following the nomenclature proposed by Holt (1960), as it pertains to a given surface area.

19. The plots in Figure 9 are all derived from the database of stock biomass and mortality time series assembled by R. A. Myers (http://fish.dal.ca/~myers/welcome.html) as included in FishBase (www.fishbase.org). The following stocks are included: cod (NAFO divisions 2J3KL, 3NO, 3PS, 3PN4RS, 4TvN, 4VsW, and 4X, W Greenland, Iceland, Faeroe Island, Celtic Sea, Irish Sea, NE Arctic, North Sea, Skagerrak, Kattegat, ICES area VIa, Baltic areas 22-24); haddock (NAFO division 4X, Faeroe Islands, NE Arctic, North Sea); hake (Northern and Southern ICES regions); saithe (NE Arctic); whiting (ICES area VIa); blue whiting (Northern ICES region); monkfish (Northern ICES region); redfish (NE Arctic); red porgy (off N. Carolina); Greenland halibut (NAFO area 3LNO, W. Greenland); and bluefin tuna (W. Atlantic; also see NRC 1994).

20. For the economic analyses here, a database was constructed of the landed values of all products from ecosystems of the North Atlantic using nominal prices from 1950, adjusted for inflation through the U.S. consumer price index (CPI), and based on constant year-2000 dollars. The information provides insights into the trends in these values over time as well as "snapshots" at particular periods (Sumaila 1999). Price information from the U.S. was used as a first approximation; country-specific figures will be included in future versions of the *Sea Around Us* database.

21. Most of the information in this section is from Tyedmers (2001).

22. There are two requirements to estimate the energy consumed. First, the catch in each fishery has to be apportioned between the types of gear and classes of vessels. Second, the current fuel consumption of the gear and vessels must be estimated. Most recent analyses of energy in fisheries have depended on data collected in the 1970s and almost every parameter has changed since then, e.g., fish abundance, fleet and vessel size, trip length, fuel efficiency, and fuel type. In some cases, actual fuel consumption data from fishing companies were obtained. From their catch data, the fuel used per metric ton of fish landed was estimated; this unit can then be used to estimate fuel use in fisheries for the same species where horsepower data are unavailable.

23. Ideally, the indirect energy that is used in fishing operations, such as boat construction and maintenance, fishing gear and labor (including the energy needed to maintain the laborers) should also be measured (Wackernagel and Rees 1996). However, fuel energy usually amounts to between 75% and 90% of the total and, given the diversity of vessels and gear, it was decided to concentrate on fuel consumption. Estimating energy consumed in fisheries is discussed in more detail in Tyedmers (2001).

24. Translation of fuel used into carbon dioxide produced follows a Lloyd's rating of 2.66 kg of gas per liter of fuel. To this is added indirect gas emissions associated with producing, transmitting, refining, distributing, and dispensing the fuel, of 0.50 kg gas per liter fuel – total 3.16 kg/liter (see Tyedmers 2001).

25. Initially, the amount of fuel as liters per metric ton of fish caught is calculated, and then converted to weight of fuel per unit weight of fish. The result is then inverted to show weight of fish per unit weight of fuel (see Tyedmers 2001).

26. Deriving protein energy production requires converting catch data on a species-specific basis into protein output, and using a formula to convert that to energy. Details are given in Tyedmers (2001).

27. More details on food sector comparisons of energy out/energy in Tyedmers (2001).

28. Another way of expressing fisheries impact on ecosystems is through the fraction of total primary production that is required to sustain the fisheries (PPR). On the continental shelves, from which about 90 percent of fisheries catches originate, the PPR is as much as 35 percent (Pauly and Christensen 1995), which is similar to the average use of primary production by humans on the planet's land areas (Vitousek et al. 1986).

29. The method used to derive these biomass maps involved the construction (by *Sea Around Us* Project staff, and a wide array of international collaborators) of 23 mass-balanced food-web models of ecosystems covering one-third of the North Atlantic, and the overwhelming bulk of its shelves. These models pertain mainly to the 1980s and 1990s, but there were enough cases from earlier periods (including the two century-old models illustrated in Figure 7) for biomass trends to become visible. These models were re-expressed in spatial form (as 1/2 degree lat./long. cells) using the Ecospace routine of the *Ecopath with Ecosim* software, also used to construct the 23 models underlying the food web models. A multiple linear regression model, comprising numerous biotic and abiotic variables (including annual catches by 1/2 degree lat./long. cells, as illustrated in Figure 2) was then used to predict the biomass of fish with trophic levels of 3.75 or more in each of the approximately 20,000 1/2-degree cells in the North Atlantic as a function of catches, water depth, etc. This procedure, described in detail by Christensen et al. (2001, 2003), differs markedly from standard method for biomass estimation, but led to results which, as shown in Figure 17, predict the same trend as the systematic application of single-species stock assessments, illustrated by Figure 16.

30. Documented in Pauly et al. (1998c). The use of declining trophic levels as an indicator of marine ecosystems was initially suspected by some of oversimplifying a complex issue in view of limitations and possible misinterpretations of the available data (more specifically, issues concerning taxonomic resolution in assigning trophic levels; use of landing statistics to reflect catches in each ecosystem; the problems caused by inclusion of some fish farming production in the landings data; and the influence of eutrophication in increasing the numbers of lower trophic-level fish). These concerns were aired (Caddy et al. 1998) and addressed (Pauly et al. 1998a) in the journal *Science*, where it was shown that the rates of decline (the trophic declines) used were, in fact, rather conservative and that consideration of the above limitations led to greater (i.e., worse) trophic declines. Later studies by Pauly and Palomares (2000, 2001) and a detailed analysis of Canadian fisheries data (Pauly et al. 2001) fully confirmed the first analysis. A number of contributions by other authors, in press at the time this is written, also confirm this.

31. Taken from an Associated Press article, which quotes *World Catch News Network* (www.worldcatch.com).

32. A formal demonstration of this may be found in Neutel et al. (2002).

33. However, several hundred thousand harp seals continue to be taken annually in Greenland and eastern Canada (Warren et al. 1997).

34. Endangered species are those listed in CITES, the Convention on International Trade in Endangered Species of Wild Fauna and Flora, and those on the IUCN Red List (Hilton-Taylor, 2000).

35. The northern right whale has recently been described as "functionally extinct"; their survival now depending on a rapid and strong reduction of the human-induced mortality (notably through entrapment in traps and other fishing gear) suffered by the adult females (Fujiwara and Caswell 2001).

36. Although commercial whaling is largely a thing of the past (with a few exceptions mentioned earlier), it is instructive to review its history in the North Atlantic both because of the lessons we can learn from the rampant and unregulated over-hunting of whales and the resulting effects on the ocean's ecosystems, which parallel those of overfishing. The following summary is based mainly on Francis (1991).

The first commercial whaling in the North Atlantic began in the Bay of Biscay as early as the eleventh century, by Basques who harpooned and lanced "right" whales—the right whales to catch—from small craft, towed them to shore stations for processing and sold nearly every part of the animals in ready markets around Europe. They probably learned of whaling from the Normans and Vikings to the north who trapped whales by herding them into shallow bays.

In the early fifteenth century, the Basques had begun whaling northward around the Atlantic coast of Europe and on to Iceland and Greenland. History does not tell us whether this "diaspora" of Basque whalers was the result of local excessive hunting, but virtually every whaling ground discovered since then throughout the world has undergone short-lived booms followed by fairly rapid declines—as a direct result of too much hunting.

By the end of the fifteenth century, the abundant fish populations off the east coast of Canada had become known to European fishers, and Basque whalers were not long in exploring the whaling potential of Newfoundland, which they exhausted by the late 1500s. (The whalers found abundant right and bowhead whales in the Strait of Belle Isle, set up stations in a bay off the strait and ran a lucrative summer whale fishery from the mid- to late-sixteenth century, when declining whale numbers, as well as the effects of European warfare [most of the Spanish armada had sunk in 1588] forced its closure). A few whalers moved into the Gulf of St Lawrence; some others turned to the fur trade and cod fishery.

In 1607, the British discovered an abundance of whales during explorations near newly discovered Spitsbergen. The British had no history of whaling but, sensing profits, arranged for Basque experts to join them. The whales were bowheads and were apparently so dense that they bumped against cables and the boats themselves. The news quickly attracted whalers of other nations, principally the Dutch. Although the British tried to enforce a monopoly, the Dutch eventually made their point by force of arms and an agreement was reached in 1618 between the two nations to base themselves in different harbors and bays on the island. Later Danish, French, and German whalers arrived and used various other harbors there. Thousands of Dutch whalers and artisans occupied the island in summer at the height of the fishery in the 1630s, but by the 1640s the numbers of bowheads were clearly in decline and by 1670 the area was completely deserted.

Whaling boats (they were nearly all Dutch by then) were forced to look farther offshore, leading to on-vessel processing (boiling blubber for oil or slicing it thinly into barrels, and removing and cleaning the baleen). This development enabled vast new bowhead grounds to be established between Spitsbergen and Greenland. By 1720 the Dutch found another concentration of bowheads in the

Davis Strait on the eastern side of Greenland. The British, who by 1670 had only one whaling vessel operating, became interested and by subsidizing whaling vessels began to compete with the Dutch in the Davis Strait in the 1750s. By the end of the eighteenth century the British had become the dominant Arctic whalers. The numbers of their principal prey, the bowhead whales, began to decline in the 1830s, and the industry was kept afloat by developments in technology—the advent of steam power and powered harpoons—the establishment of shore stations in Davis Strait, seal hunting and, from the late 1800s, a spectacular rise in the value of baleen for shaping women's garments. By the first decade of the twentieth century, bowheads were rare in the Davis Strait, and whaling there ended shortly before the beginning of the First World War. (On the Pacific or western Arctic side, a similar story was played out by American whalers chasing bowhead whales. Whaling petered out there in the first decade of the twentieth century, leaving a few thousand animals, a slightly better picture than in the eastern Arctic, where only a few hundred bowheads remained.)

American colonists began hunting right whales from small shore-based boats in the 1640s, first at Long Island and later Nantucket. In the early 1700s, Nantucket whalers began to hunt sperm whales farther offshore, and by the 1730s had spread north to join whaling vessels out of England in the Davis Strait and Newfoundland. The restrictions placed on the colonial whalers were such that they spread across the South Atlantic (and out of this history). New England whaling ceased for three years during the American Revolution but began to bloom again when American whalers again ventured south along both sides of the Atlantic and then discovered the rich grounds of the Pacific Ocean in 1789. Some Nantucket whaling companies moved across the Atlantic to Britain and France while others moved north to (British) Nova Scotia.

The 1860s ushered in the modern age of whaling with the Norwegian development of the steam-powered, steel catcher vessel and more efficient harpooning equipment. The initial targets were rorquals—humpbacks, finbacks, blue and sei whales—that were too fast and elusive for the traditional whaling methods based on small, rowed wooden whaling boats. Schools of these baleen whales passed along the Norwegian coast between their northern feeding grounds and temperate waters to the south where they calved.

Twenty years after the launching of the first steam whaler in 1863, there were 23 vessels catching and towing whales to shore stations in Norway. By then, the whale numbers had already declined. The Norwegian whaling industry consequently spread to Iceland (where the grounds were overfished and whaling banned by 1913) and the Faeroe Islands, reaching Newfoundland across the Atlantic in the 1890s (where, after an initial boom in the early 1890s that resulted in 27 shore stations there by 1904, the industry gradually collapsed; only three stations remained by 1915), and later to the Shetland Islands and Spitsbergen, and indeed in the early twentieth century to whaling grounds around the world as far south as the Antarctic.

Norway pioneered whaling regulation with the 10-year ban on whaling off Finnmark from 1904, and in 1929 by a Whaling Act that restricted the species that could be killed and set minimum lengths for them.

International concern was shown in the adoption of a Geneva Convention in 1931 that introduced certain agreed restrictions, which, however, were not linked to penalties.

The International Whaling Commission, formed to deal with new whaling industries that developed after World War II, first met in 1949. It largely failed to control whaling until a necessary majority of nations in the commission agreed to a moratorium on whaling worldwide that took effect in 1986. Aboriginal hunting in various parts of the world was allowed to continue, not without emotional debate, and loopholes have allowed some commercial-level "scientific" whaling by Iceland, Japan, Norway, and Korea.

However, it was not scientific debate that "saved the whale" but, importantly, a groundswell of public opinion that the whale was "one of those animals which for man have a special significance," as Sir Sydney Frost said in summing up in 1979 the results of a national inquiry into the future of whaling in Australia. The moratorium was to have been revisited by the IWC in 1991, but for the foreseeable future there is no inclination by the international community in general to recommence whale hunting. Nevertheless, the International Whaling Commission has in place a revised management procedure that (in 1993 at least) was "in all likelihood the most tested procedure in the history of resource management" (Magnusson 1993), which would swing into place should commercial whaling ever resume.

37. Kaschner et al. (2001), based on a methodology earlier presented by Trites et al. (1997), estimated marine mammal food requirements in the North Atlantic from estimates of mean body sizes, feeding rates, and diet composition in eight food categories. These were linked to GIS-digitized distribution maps that included refinements for habitat preference. Resource overlaps with fisheries were calculated in relation to fisheries catches and food requirements of marine mammals in the same 1/2 degree spatial cell.

38. These measures and data on whale watching are included in the marine mammal section of the World Wildlife Fund's Internet site (www.panda.org).

39. An example is the Benguela upwelling system, which has a different structure during cold upwellings and stable warm conditions (Moloney 1999).

40. It has now been demonstrated (Jackson et al. 2001) that pristine biomasses of large marine organisms were in ancient times much higher than would be suggested by examining data from even the earliest period of industrial fishing. Exploitation of marine mammals and other large marine organisms thousands of years ago generated long-term impact on the abundance of non-target species; impact that is still being felt today.

41. See for example McGoodwin (1990).

42. There is now convincing historical evidence (Jackson et al. 2001) that overfishing, which removes the inhabitants of various ecological niches, is a prerequisite for such factors as organic pollution and habitat modification to degrade marine ecosystems. Healthy ecosystems are much more resilient to such impacts.

43. Odum's maturity scale for aquatic ecosystems was examined by Christensen (1995a), Christensen and Pauly (1998), Vasconcellos et al. (1997) and

Christensen and Walters (2000), and found to be compatible with the response of aquatic ecosystems to fisheries and to releases from this stress.

CHAPTER 3

1. There is a huge literature on "perverse incentives" (i.e., side effects of regulatory action resulting in effects opposite to those that were intended); much of it is accessible via a straightforward web search.

2. Weber (2002).

3. Weber (2002); Pauly (1996) gives a history, and the scientific background (if any) of these estimates.

4. Weber (2002).

5. Daan (1997).

6. The information and discussion in this section are drawn mainly from Munro and Sumaila (2001).

7. In recent years, these include the World Trade Organization, the Food and Agriculture Organization of the United Nations (FAO), the Organization for Economic Co-operation and Development (OECD), and the United Nations Environment Programme (Porter 2001).

8. More programs with similar effects may be cited, notably the U.S. "Fisheries Obligation Guarantee Program" described in Weber (2002), and a number of equivalent programs in other countries around the North Atlantic.

9. Munro and Sumaila (2001) cite two studies on the size of fishing subsidies, the lower range from a 1998 study for the World Bank, and the higher value from a 1992 FAO study.

10. Details are given by individual European countries and by proportion of total national catches in Munro and Sumaila (2001), based on the Organization for Economic Co-operation and Development (OECD) data; the subsidies are shown in the eight categories described above. At least one observer believes that the relatively high proportions of subsidies for Canada and the U.S. may be due to underreporting of subsidies by other countries (Dr. A. Rosenberg, University of New Hampshire, personal communication, 2001).

11. A prominent example is the expansion of the U.S. fishing fleet in the late 1970s and 1980s following declaration of the exclusive economic zone. Two government-funded programs—The Fishermen's Capital Construction Fund and the implementation of the American Fisheries Promotion Act—led to overcapitalization in the fisheries (Hall-Arber and Finlayson 1997).

12. See Kaczynski and Fluharty (2002). The issue of European export of excess fishing capacity to overfished areas of West Africa is followed in documents assembled to support a symposium held in Dakar, Senegal, in June 2002 (see www.saup.fisheries.ubc.ca /Dakar/index.htm.)

13. Mentioned in Porter (2001).

14. A review of the EU's Common Fisheries Policy in 1992 showed the need to restructure the whole sector, in part to remove excess fishing capacity. There

are four main channels through which assistance (subsidies) is given: the Financial Instrument for Fisheries Guidance, which provides substantial support for infrastructure development, vessel renewal, vessel scrapping, etc; the European Regional Development Fund, which provides general infrastructure, etc. support to disadvantaged regions of which a small part is used for fisheries; the European Agriculture Guidance and Guarantee Fund, from which a small proportion is used for price support in favor of fishers; and the European Social Fund, which mainly supports persons to become more employable. In addition, the PESCA initiative from 1994 to 1999 assisted fishers to diversify away from fisheries and developed new employment opportunities. Also, fisheries agreements between the EU and different countries include payments for fishing rights and assistance to local fisheries, research and monitoring, and control. Finally, there are special funds for research and for inspection and monitoring. (This summary adapted from Anon. 2001b.)

15. The serious shortfall in EU capacity reduction exercises is summarized from Porter (2001). Not mentioned there are the enormous costs that must be involved in the scheme.

16. A persuasive economic model is given in Munro and Sumaila (2001).

17. These subsidies took different forms. Unemployment insurance (later called employment insurance) benefits due to the seasonal nature of fisheries in many coastal communities made up on average 30% of the income of fishers in these communities in Atlantic Canada in 1990. The government set up an Atlantic Groundfish Strategy of income support for fishers and shore workers in coastal communities displaced by the 1992 cod fishery closure with a budget of Can$1.9 billion. Also the government provided subsidies for boat building associated with the introduction of exclusive economic zones (EEZ) in 1977 and restrictions on fishing companies in the early 1980s (Greer 1995; de Young et al. 1999). Subsidies by Canada to its offshore fishing fleet from 1954 to 1970 increased its capacity by a factor of 18, which the government later acknowledged contributed to overcapitalization in the fishery. A new round of subsidies following the declaration of EEZs increased fleet capacity to five times that needed to catch the annual cod quotas; both increases were followed by collapses of the cod populations (Porter 2001).

18. The detailed arguments are given by Munro and Sumaila (2001).

19. Information in this paragraph from Sumaila et al. (2001).

20. Details are in Ruttan et al. (2000).

21. Taxpayers do not always pay for fisheries management; in Australia and New Zealand, the fishing industry pays for these services through fees and other charges (Porter 2001); the problem, then, is that the industry determines what it does not want researched, which makes it unlikely for critical assessment of its performance to emerge.

22. OECD (2000).

23. Arnason et al. (2000).

24. Details in Schrank and Skoda (2001).

25. This description comes mainly from the Independent World Commission on the Oceans (1998).

26. From Hall-Arber and Finlayson (1997). This situation provided a dilemma for the Canadian Government because the banning of the new vessels would have caused significant economic disruption. The result was the institution of individual transferable quotas in 1989 for the new vessel owners in the inshore fleet. These remain a source of controversy between them and the fixed-gear inshore fishers.

27. OECD (1997).

28. NRC (1999b).

29. Weber, *From Abundance to Scarcity* (2002). Weber focuses on the often conflicting mandates of the U.S. National Marine Fisheries Service (NMFS) and its dual role of serving the industry and protecting the fisheries. As Weber notes, "Indeed, it was as if the NMFS were really two separate agencies: one that created fishing pressure through marketing, subsidies, and technical assistance, and another agency that tried to contain that pressure through regulations." The result is a downward spiral in which one fishery after another is first exploited and then, too late, protected, before it finally collapses. Weber also describes in detail how the fishing industry dominates the regional fishery management councils, which create the management plans that are supposed to control overfishing. Not surprisingly, these plans have usually proven inadequate to spur recovery of stocks.

30. Weber (2002).

31. From an article based on the UK *Independent* of 2 December 2000, and *New Scientist*, 1 December 2000, on the Internet site www.eces.org/articles/static/97560720019032.shtml.

32. Information related to ICES and the North Sea is mainly from Went (1972).

33. These and other international instruments and their mode of operation are documented in Alder et al. (2001). Maguire (2001) gives an account of how ICES and NAFO generate formal "management advice."

34. Criteria and source of information used to evaluate compliance with international fisheries conventions are detailed in Alder et al. (2001), from which this section is adapted.

35. The other relevant international instruments, which are of limited relevance or superseded, are shown below (from Alder et al. (2001). (See table next page.)

36. Description adapted from that institution's website (www.nasco.org.uk).

37. See NEAFC (1985, 1999).

38. Canada and Iceland withdrew from IWC; Iceland then rejoined, while Canada maintains a high level of compliance, though it allows hunting by First Nations (e.g., Inuits); Finland and Greenland also allow aboriginal whaling.

39. See Chuenpagdee and Alder (2001).

40. Our social activities are regulated explicitly or implicitly by institutions. Hence our interactions with the resources of the North Atlantic Ocean are presently

INSTRUMENT	NOTES
Superseded	
Convention on the Territorial Sea and the Contiguous Zone	Replaced by UNCLOS
Convention on Fishing and Conservation of the Living Resources of the High Seas	Replaced by UNCLOS
Convention on the High Seas	Replaced by UNCLOS
Convention for the Regulation of the Meshes of Fishing Nets and the Size Limits of Fish	Replaced by Northeast Atlantic Fisheries Convention
Fisheries Convention	Replaced by UNCLOS and CFP
Agreement on the Regulation of North-East Atlantic Cod	Replaced by Cooperation in the Area of Fisheries
Reciprocal Fisheries Agreement	Replaced by NASCO
Agreement on the Regulation of Fishing of the Atlantic Scandinavian herring (1973)	Replaced by the 1980 Herring Agreement
Agreement on Fisheries and the Continental Shelf Between Norway and Iceland (1980)	Replaced by 1996 and 1997 Herring Agreement
Limited Relevance	
Declarations from International Conferences on the Protection of the North Sea	Ministers from the North Sea reconfirmed their commitment to improving the state of North Sea resources through cooperation with existing structures. No actions or initiatives initiated.
Convention on the Continental Shelf	Focused on sedentary species and the seabed.
US-Canada Agreement on Fisheries Enforcement	Mutual cooperation in setting enforcement standards and implementing policies.
Convention on Conduct of Fishing Operations in the North Atlantic	Focused on gear and associated equipment, it is implemented by other agreements where gear or other equipment is specified.
Agreement on Sealing and the Conservation of Seal Stocks in the Northwest Atlantic (amended 12–Dec-75)	Bilateral between Canada & Norway, does not appear to be used today by either party, each party sets own quota.

continues

INSTRUMENT	NOTES
Limited Relevance (continued)	
Agreement on the Measures to Regulate Sealing and to Protect Seal Stocks in the Northeastern Part of the Atlantic Ocean	Bilateral between Norway & Russian Federation, appears to be used to set annual quota for harp and hooded seals.
Agreement on Cooperation in Research, Conservation and Management of Marine Mammals in the North Atlantic	Greenland; Iceland; Norway; Faeroe Islands.
Convention on the International Maritime Organization	Conduct of vessel operations, no role in fisheries management.
Agreement Concerning Cooperation in Marine Fishing	Signed in 1963 between ex-USSR, Cuba, Poland, and Bulgaria. Its currency is unknown. Focused on open seas and the development of new fisheries.
North American Agreement on Environmental Cooperation	Involves Canada and the USA; no commercially important marine fish are included.
Agreement on Fisheries Between Norway and the Faeroes	Focused on giving mutual rights to fish in each country's EEZ, does not manage shared populations.
Convention on International Trade in Endangered Species of Wild Fauna and Flora	Does not include commercially important marine fish species.
Convention on the Conservation of Migratory Species of Wild Animals	Does not include commercially important marine fish species.
Convention on the Conservation of European Wildlife and Natural Habitats	Does not include commercially important marine fish species.

controlled by a given set of institutions, and any alternative management measures would similarly be governed by institutions. Institutions are rules, established laws, customs or practices, and important relationships (e.g., marriage). In fisheries, they include such entities as the market, fishers' cooperatives, and local governments, and even "working knowledge" of fishers that may or may not closely resemble formal laws and court decisions (Hall-Harber and Finlayson 1997). Organizations, with which they are often confused, are sometimes physical embodiments of institutions, or simply the structure of, say, a business or a political party.

Viewed from this perspective, developing countries and the marginal regions of some more developed countries that have taken the route of local community management have preempted this trend in the West, and can offer guidance on the types of institutions that may help devolve management to appropriate local levels. Of course, in most cases in developing countries, this type of management is simply a revival of the traditional way of doing things, which was supplanted by government proclamations as a part of colonization processes in the past. An excellent example is Apo Island marine reserve in the Philippines. There, community involvement was the sine qua non of its success (Russ and Alcala 1999).

The fact that so few examples of successful co-management are reported in the current North Atlantic fisheries literature implies that this is a difficult path to tread. Yet, this is the mode urged in that same literature. A co-management model for European fisheries that includes elements of government, market, and civil society is seen as "more than an option, it is a necessity." (Kooiman et al. 1999).

A community-based approach does not seem applicable to large-scale fleets that are often company-based, sometimes multinational, and that may have no "home" community, province, state, or even country. It may be that there are faults in the present approach to regulation and management of these large-scale fleets, and that means could be sought to change the nature or reduce the scale of their operations to fit, along with other fisheries subsectors, into the purview of more local authorities. This would allow the better inclusion of the views of small-scale fishers, who at present usually have less access to the political arena.

Reduction in the scale of fishing operations also has to be tempered by our present expectations: some of us want to be able to find frozen fillets in the same place in the supermarket shelves each week; fast food outlets need to be assured of a continuous supply of fillets of a specified fish. Both society at large and fishers need to compromise all along the way. Further, because North Atlantic fisheries encompass a wide spectrum of fishing conditions, fishing communities and societies, different compromises to suit different situations are inevitable.

The Committee on Sustainable Fisheries of the U.S. National Research Council has put forward a "virtual" community approach: instead of management by a "place-based" community, the proposal is for management of a resource by its functional equivalent (NRC 1999b). In this approach, management responsibility would be devolved to those fishing a resource, regardless of geographic location—the virtual community. The problems of "stateless" international fishing companies might be overcome through this approach. However, a different but major problem arises: the consequent mixing of large companies and individual vessel operators would likely end in the former squeezing out the latter. Real, place-based communities of small-scale fishers would suffer in such arrangements.

Regardless of the location of the fishing entities—companies and/or communities—it is argued here that the location of the fishing grounds and the ecosystem(s) that surround them is also most important. Ecosystem boundaries should be the boundaries of any such "communities," which would require, as

does any responsible institution, mutual communication, trust, shared understanding, and collective action.

From a societal point of view, a desirable outcome of future fisheries management would be a transition from fisheries being in the hands of the large-
scale sector and government to more diffuse fisheries where all resource users
have a share and largely control themselves and each other—with government
representing society by providing oversight.

In fact, it is becoming almost conventional wisdom in the literature that control of fisheries should devolve as far as possible from central government
toward institutions in fishing communities. Central government agencies are
obviously where catch-related data from national and international fisheries and
other information related to the ecosystems should continue to be collated and
analyzed and advice provided to the communities.

But can such communities, which are usually fishing communities, virtual or
not, maintain ecosystems as opposed to control fishing? Where does fisheries
co-management or fishing community management fit in this broader picture?
Where, indeed, does traditional knowledge fit when that knowledge is of fisheries or ecosystems long since depleted or despoiled and under increasing,
potentially irreversible, pressure from nontraditional fishing and other human
impacts? Who will call for moderation instead of for maximizing catches (and
the conflicts and ensuing expensive, but nearly always futile, management
processes and structures)?

41. This sentiment from McGoodwin (1990).

42. These desirable features of improved fisheries governance are expressed in
McGoodwin (1990); Cochrane (2000) also list desiderata for re-invented fisheries management procedures.

CHAPTER 4

1. Complete denial of the true situation of the state of the resource by fishers
and/or vessel owners is more frequent than one should think. Here is a typical
quote, from a Canadian seiner operator: "We're catching lots of fish—don't let
people fool you that there's no fish out there . . . " (from *Research on Business
Magazine, Globe and Mail,* Toronto, May 2001, p.144).

2. Fisheries agencies, both international and national, have routinely seen the
demand for fish as something that must be met. The world (the nation) must
produce so many more hundreds of thousands of metric tons in order to meet
projected world (domestic) demand in 2025. This is an entirely circular statement, because it is not demand but supply that is the basis of the projections.
What is meant, really, is the amount required to maintain present per capita fish
consumption. If the oceans were less bountiful and supply had reached a crisis
similar to the present one at only half its present amount, then the projections
would be for a correspondingly smaller "demand." The "demand" for fish in a
developing country such as the Philippines is really a question of income. If
Filipinos had more income they would buy more fish (Posadas et al. 1986). In

more developed countries, consumption of fish has to some extent become alternatively a matter of taste, fashion, or health, much of it driven by the media.

This vicious cycle of demand based on production and consequent exhortations and/or incentives for production to meet demand have to be broken. Neither the ecological, nor the economic components of fisheries science support this kind of cycle. It may still apply to agriculture, where there is still "unutilized" land (as agriculturists call pristine ecosystems) to be plowed or yields in existing farmland to be increased, and to the mining industry, as long as new seams and oil-bearing strata can be found. However, the time has come to remove "demand" as the basis for deciding on future fisheries investments.

As described in Sobel (1975), the roots of the issue lie in the world food crisis of the early 1970s when the bubble of postwar food complacency burst. It was a time of failing crops—there was a severe El Niño event in 1972–1973; spreading drought in Africa; starvation in South Asia amid the turmoil associated with the birth of Bangladesh; the first major oil crisis, which sharply increased fuel prices; and some countries did not or could not honor food export contracts, escalating the cold-war tension between the then "superpowers" and between them and China. By mid-1974 the world's grain reserves were only enough to last 26 days. Norman Borlaug, the Nobel Peace Prize winner who led development of the high-yielding grains that began the Green Revolution of the 1960s, predicted that 20 million persons might die within a year.

The seriousness of the situation can also be seen in some of the more extreme views of the time on population, such as application of the triage concept and "lifeboat ethics." Triage (from the French *trier*, to sort) was the medical practice of separating the wounded during World War I into three groups: those who would die regardless of treatment; those who would survive without treatment; and those who would survive only with treatment. Under circumstances of limited medical personnel, assistance was only given to the last mentioned group. Similarly, it was argued, food aid should only be given to those countries whose population growth had not exceeded the potential of their agriculture. Giving food to nations that cannot feed themselves "is to throw sand in the ocean." Garrett Hardin, well known (or notorious, given one's viewpoint) for his work on "the tragedy of the commons" (Hardin 1968) compared the rich nations to people in a lifeboat while the poor nations were those in the sea swimming around them. Allowing the swimmers into the boat would cause it to sink; even admitting a few would mean the loss of a necessary "safety factor." Sharing food with poor nations would, he felt, be likewise foolhardy.

There was corresponding panic in the fisheries sector. After spectacular postwar growth in world catches, there was a sudden and unprecedented decline in 1969 and although growth continued in 1970, concern remained. The largest fishery in the world, that for the Peruvian anchovy, collapsed in 1972. The then-former U.S. Secretary of Agriculture, Orville Freeman, pointed out to a U.S. Senate Agricultural Subcommittee in 1974 that "many marine biologists now feel that the global catch of table-grade fish is at or near the maximum sustainable level." Soviet trawlers were ramming Canadian fishing boats and destroying the gear of U.S. lobster boats; "cod wars" broke out again (the first was in the late 1950s) when British, Belgian, and West German trawlers defied Iceland's further extension of its exclusive fishing zone. Countries one after the other were asserting extended fishing rights, some proclaiming 200-mile exclusive economic zones. Latin American countries were seizing U.S. fishing boats entering their new territories.

In response to the food crisis, a World Food Conference (and a World Population Conference) was held in 1974. The food conference expressed "concern at the inadequate performance of agriculture, including livestock and fisheries, in many developing countries in relation to . . . targets," called for *high* (emphasis in the conference resolution) priority for agricultural and fisheries development, for research to develop fisheries, and for assistance to developing countries to develop their fisheries.

"Inadequate performance" in relation to meeting targets is hardly a statement designed to provide a sense of stewardship of marine resources. It reflected the overreaction of many countries to the crisis of the early 1970s just as clearly as did the triage and lifeboat analogies. Yet, the notion carries over to the present, a sense that fisheries ought to "perform" better in order to reach targets.

Preparatory meetings toward the UN Convention on the Law of the Sea (UNCLOS) were already underway during the 1970s food crisis, and the text of the convention in 1982 relating to fisheries echoes the concerns of that period: "The coastal state shall determine its capacity to harvest the living resources of the exclusive economic zone. Where the coastal state does not have the capacity to harvest the entire allowable catch, it shall, through agreements . . . give other states access to the surplus of the allowable catch . . . " (Article 62). One way or another states became encouraged, obliged in fact, to boost their fisheries.

The UN Conference on Environment and Development in 1992 does specify a number of conservation measures to be undertaken toward conservation and sustainable use of marine living resources under national jurisdiction, such as using selective fishing gear; protecting and restoring endangered species; preserving rare or fragile ecosystems, habitats, and other ecologically sensitive areas; and maintaining populations of marine species at levels that can produce maximum sustainable yields. But the sense of taking to the limit, sharing or not, remains. States are requested to "Develop and increase the potential of marine living resources to meet human nutritional needs, as well as social, economic and development goals" (Agenda 21).

These concepts are carried over into the FAO Code of Conduct for Responsible Fisheries, which, although the caveats therein are more detailed and stronger, repeats the message that adopting states should maintain fish populations at levels capable of producing maximum sustainable yields and promote their optimum utilization. "Optimum yield" is defined in the Magnuson-Stevens Act that is the authority for U.S. fisheries, as "the maximum sustainable yield from each fishery, as reduced by any relevant economic, social or ecological factor." One recent recommendation, the U.S. National Coalition for Marine Conservation (Hinman 2000) is to recast that definition with more explicit obligations to consider ecological factors, including predator-prey relationships.

The idea of *moderation*, however, as opposed to maximum use is nowhere to be found. Neither can we call the "precautionary" approach a tool for moderation; it is simply a way to approach the maximum sustainable (or optimum) yield without overshooting. All the prescriptions for sustainable fisheries push us toward the cliff edge of management in a fog of inexact information, instead of following a route on safer ground.

3. There are some examples of a tentative transition toward ecosystem-based governance which can be used as guides. These can be found in the reports of the Commission for the Conservation of Antarctic Living Marine Resources, the Ministerial Declaration for the Black Sea, and the Ministerial Declaration for the North Sea. The Barents Sea ecosystem is the subject of another effort in this regard (Sherman et al. 1996).

   Ecosystem management obviously involves more than just regulating the fisheries sector, particularly in coastal ecosystems where all manner of enterprise, shipping, and water-use sectors intersect and compete for use of ecosystem resources. The intent of the concept is to place ecosystem integrity or health as the primary consideration in the management of such areas. The inestimable value of their ecological functions and services is justification enough.

4. A more accurate description of the present direction for medium-term management is "fisheries-based ecosystem management." In the long term, of course, the goal should be simply "marine ecosystem management," meaning maintaining ecosystem integrity in a sustainable manner and all that it implies, on the one hand, for fishing (and other activities that potentially harm marine ecosystems) and, on the other, for the continued functioning of and provision of services from the ocean.

5. Yodzis (2001).

6. Protocol for the Scientific Evaluation of Proposals to cull Marine Animals. (www.cull.org).

7. The degree of direct competition between marine mammals and fisheries is overall quite low, as we have shown. There are definite "hot spots" of high overlap, but fishers also tend to use marine mammals as scapegoats for their own excesses. Some perspective is needed here. Remember that while marine mammals in the North Atlantic consume about as much commercially exploited fish as taken by the fisheries, other fish consume several times what the fisheries take. Indeed, the fate of the "average fish" is to be consumed by another fish.

   In any case, the statement about "dwindling fish stocks" makes the competition aspect (the rationale for culling) glaringly redundant. Culling would only slow down the dwindling—and then only until the fishers take up the slack. The real problem to be addressed is overfishing.

8. An approach exists that might help not only to turn the tide, but also to gather community support for a new era of fisheries—essential because there can never be enough police to patrol the coasts and ocean. We call this process "Back to the Future" (BTF), and it is briefly outlined here.

   BTF is based on the realization that present fisheries management policy of trying to ensure sustainability has to be misguided in an ocean as depleted of large, valuable fish, and as impoverished of its pristine diversity of birds and mammals, as we see today. Indeed, this book suggests things are getting worse more rapidly than we feared. In the face of these ratchet-like processes (Pitcher 2001a), the only viable policy is to rebuild resources, as outlined above.

If this logic is accepted, we need a rebuilding policy goal against which we can measure progress. In BTF this goal is an ecosystem that existed at some time in the past. Note that policy goals need not ever be reached. They are set as an aim, for managers, fishers, and the public to gauge progress and check if various measures are actually working as hoped. Moreover, the best of policy goals act as an inspiration, the very nobility of their nature encouraging compliance and consent. For our rebuilding policy we have to seek an ocean ethic that fits with Aldo Leopold's Land Ethic: "A thing is right when it tends to preserve the integrity, stability and beauty of the biotic community" (Leopold 1949).

The best policy goals also have a day-to-day practical side: they encourage practitioners and the public to check each small action to see if it helps or hinders. Not many fishery policy goals fall into this category, although maybe the "maximum sustainable yield" of the 1960s, likened by Larkin (1977) to a "stained glass cathedral" came close.

Restoring an ecosystem that existed at some point in the past meets all these criteria. Since it actually existed, the animals and plants must be in balance in the sense of having enough food (and habitat). Management actions (and in fact every action by every fisher) can be judged to see if they help or hinder. In practical terms, we seek to build model reconstructions of the past (see Christensen 2001). To do this we can use archeology, data archives, and historical documents. Traditional and local environmental knowledge, including oral tradition, can also play an important part (see contributions in Pauly et al. 1998b). A number of snapshots of past ecosystems can then be compared to see which is the best goal to adopt. This BTF policy work is still in its infancy (Pitcher and Pauly 1998, and Pitcher et al. 1999 provide early examples), but does present a useful vision of how the present, depressing trends may be reversed.

9. Inshore habitats are nearly all affected also by pollution, and there is a huge literature to document this. In this book, we concentrate on fisheries impacts, as these tend to be overlooked in public forums, where overfishing is perceived mainly as impacting on fishers' income, and market supply.

10. This view from the (U.S.) National Research Council (NRC 1999b). A good picture of the present pain both to scientists and fishers in the declining Gulf of Maine fisheries is sketched in *The Great Gulf* by David Dobbs (2000).

11. Such a scenario has been discussed—if somehow tongue in cheek—for the North Sea by Pope (1989).

12. It has been proposed (ADB 2001, p. 21) that taxing fishing enterprises on the basis of their environmental resource consumption might be used to reduce this kind of fishing effort. However, the research that would be required for such a scheme to be applied equitably to, say, bottom trawling, has not been conducted, contrary to that required to implement, e.g., carbon taxes, presented further below.

13. See Anon. (2001a).

14. Information on individual quotas is in part from the (U.S.) National Research Council (NRC 1999a).

15. A critical review of the "stewardship" concept is given in Roach (2000).

16. From an extensive critique of ITQs in Greer (1995).

17. These interpretations from the (U.S.) National Research Council (NRC 1999a). The council recommended to Congress that it lift the 1996 moratorium and leave the setting of individual quotas to regional councils.

18. The need to place fisheries in their ecosystem context and the placing of humans in marine ecosystems as predators, for example, are contained in a report of the (U.S.) National Research Council (NRC 1999b). The FAO Code of Conduct for Responsible Fisheries and FAO's International Plans of Action (e.g., FAO 1999b) contain much useful advice on managing fisheries and fishing capacity (although they do not take the explicit ecosystem approach expressed here). A Reykjavik Conference on Responsible Fisheries in the Marine Ecosystem in October 2001, organized jointly by FAO and the Iceland Government, indicates an international movement toward ecosystem-based fisheries management (Conference details on www.refisheries2001.org).

19. These findings were reported in the international scientific journal *Nature* of 16 December 1999 (Jones et al. 1999; Swearer et al. 1999).

20. Russ and Alcala (1999).

21. See Guénette et al. (2000).

22. Hutchings (2000).

23. Information in this section is from Murawski et al. (2000). Factors they see as crucial to the efficacy of closed areas include the degree of fish movement across the boundaries; the distribution and quantity of displaced fishing effort; the relative catchability of the target species outside the closed areas; and the level of protection afforded to undersized animals taken by the fishery. Note also that there were "important costs in foregone fishing opportunities for the fleets."

24. Roberts et al. (2001).

25. See Roberts (1999).

26. The marine reserve network proposal was based in part on a survey by the U.S. National Center for Ecological Analysis and Synthesis. In May 2000, then President Clinton issued an executive order to develop a network of marine reserves in the U.S. The new administration is seeking to stop or delay its implementation. The use of marine reserves in general was a recommendation of the (U.S.) National Research Council (NRC 1999b), which also noted that attempts to introduce marine reserves there in the past had met with strong opposition from fishers, and that broad involvement of fishers was a key strategy in their introduction.

27. Bellwood and Hughes (2001), who review factors influencing biodiversity on coral reefs. Habitat area is far and away the most important determinant.

28. Cited in the Independent World Commission on the Oceans (1998).

29. See Roberts (1999).

30. As Walters (1998) shows, the situation is different in the Pacific, at least along the Pacific coast of Canada, where the default is that the salmon fisheries are "closed" except during fishing seasons that are set annually; this contrasts with the situation prevailing with most other fisheries, whose default setting is to be open, while closed seasons must be set annually. This may be one reason why these fisheries are still operative (though not in the best of states), while the salmon fisheries in the Atlantic are essentially gone.

31. As described in de Young et al. (1999).

32. See Zarrilli et al. (1997). This volume has 27 chapters/studies, none of which mention fish, which is clearly a new entrant in the eco-labeling field. The goals mentioned in this paragraph are from the same authors.

33. The Clean Development Mechanism, an outcome of the United Nations Framework Convention on Climate Change, helps address issues of climate change on a project basis. A Kyoto Protocol to the UN Framework Convention on Climate Change was formally adopted by the third conference of the parties on 11 December 1997 in Kyoto, Japan. It establishes legally binding obligations on signatory countries to reduce emissions of six greenhouse gases in total by about 5 percent below 1990 levels by 2008–2012. The conference established the Clean Development Mechanism (CDM), which inter alia, allows project-based trading between developed and developing countries in reducing emissions: developing countries can gain financial credits through certified emission reductions. The (signatory) developed country providing the financial credit earns credits toward its own target of emission reduction. This is similar to the idea of emissions trading, which is a market-based instrument aimed at providing flexibility and choice for achieving the most cost-effective compliance. However, the CDM is not an open-market concept; supervision would be centralized under an executive board. Information from www.igc.org/wri/ffi/climate/cbf-cdm2.htm.

34. Greenpeace is an international nongovernmental organization based in Amsterdam, with a presence in 41 countries. It describes its role as global environmental campaigning, and organizes public campaigns for, among other things, protection of the oceans. Information from the Greenpeace Internet site.

35. The Marine Stewardship Council (MSC) is an independent nongovernmental organization representing a partnership of the World Wide Fund for Nature (WWF) and Unilever. Its purpose is to establish, on the basis of a broad set of standards for sustainable fishing, a system of certification for individual fisheries that are considered to follow good practice. Seafood companies are encouraged to form buyers' groups and buy only fish products marked with a logo indicating that they have come from certified fisheries. Independent, accredited certifying firms certify only fisheries meeting the standards (WWF documents cited by the Independent World Commission on the Oceans) (1998). See also www.msc.org. However, MSC has been less than transparent in some recent decisions, such as "a blanket approval of all species, localities and gear types for Alaskan salmon fisheries" (Pitcher 2001b).

36. From Alverson et al. (1994).

37. There are three broad types of eco-labeling: voluntary labeling established by third parties including government and nongovernmental organizations for processes (such as good fishing practices); labels for a single attribute, such as energy efficiency (and which we propose should relate to the method of capture of fish products); and labels that give qualified information on an agreed set of indices, like the nutritional labeling becoming common in processed foods (Zarrilli et al. 1997).

38. Concerns about eco-labeling of fish products were expressed at the meeting of the FAO Committee on Fisheries (COFI), 26 February–2 March 2001. "The Committee was informed of eco-labeling schemes in some countries, such as in the Nordic countries and Japan. . . . Some members expressed deep concern that a private initiative such as this could become an additional barrier to trade especially if it were not based on scientific and objective criteria." A COFI subcommittee is to investigate catch certification schemes used by some regional fishery management bodies. (From the report of the 24th Session of the Committee on Fisheries, FAO, Rome, 26 February–2 March 2001.)

39. Lee (2000).

40. Behind the Scenes. *National Geographic* magazine, January 2001.

41. From Wilkinson (2001).

42. North Atlantic whale-watching data were adapted from a world review by Hoyt (2001); details are given in Kaschner et al. (2001).

43. The example of applying intergenerational equity is from Sumaila (2001).

44. Drawn from the Independent World Commission on the Oceans (1998).

45. These plans are in FAO (1999b).

46. This plan was approved by FAO's Committee on Fisheries on 2 March 2001.

47. The scientific name of Patagonian toothfish is *Dissostichus eleginoides*, Smitt 1898; the species belongs to the family Nototheniidae, or cod icefishes.

48. www.isofish.org.au.

49. www.fiskeridir.no/english/pages/list.html.

50. The discard problem is now being addressed by an ICES Study Group on Discard and By-catch Information, which will provide solid estimates of discards and/or bycatch for those fisheries currently being monitored and enable ICES to have accurate information on total catches of the respective populations. The study group first met in March 2000 and resulted in an inventory of all projects that collected discard or bycatch information in the ICES area (ICES document CM 2000/ACFM:11). FAO (2000a) also addresses this issue.

51. The first was an FAO/Australian Government meeting in May 2000, which drafted a preliminary International Plan of Action of illegal, unreported, and unregulated fishing, followed by an FAO technical consultation to finalize the plan (October 2000; FAO 2000b), and a joint International Maritime Organization/FAO working group (also October 2000; FAO 2001).

52. The International Plan of Action to Prevent, Deter and Eliminate Illegal, Unreported and Unregulated Fishing was approved by the FAO Committee on Fisheries in March 2001 and was expected to be adopted by the FAO Council (FAO's governing body) in June 2001. It defines "illegal" as fishing activities that are conducted without the permission of the State or in contravention of its laws and regulations; in contravention of measures adopted by a regional or international organization to which a State is party; or are in violation of national laws or international obligations. "Unreported" fishing is defined as unreported or misreported in contravention of the State's laws or regulations or in contravention of the reporting procedures of relevant regional fisheries management organizations. "Unregulated" fishing is defined as activities that are conducted in the area of application on a relevant fisheries management organization by vessels without nationality, or by those flying the flag of a State not a party to that organization, in a way not consistent or contravening the management measures of that organization; or in areas where there are no applicable management measures and where the activities are conducted in a way not consistent with relevant State responsibilities under international law.

53. This perception is documented in Pauly (1996b), in form of a graph drawn in the early 1970s, illustrating what was then perceived as a historical "evolution" of fishing boats, ranging from the small rowed boats of ancient times, and leading via various intermediated forms to the large factory trawlers forming the bulk of the Soviet, German, and other distant-water fleets (DWFs), then on top of marine food chains. As it turned out, these notions not only had no predictive value (the exclusive economic zones [EEZs] declared later in that decade quickly rendered DWFs outdated), but were particularly pernicious when applied to developing countries, whose small-scale fishers could never have been converted into crew on factory trawlers.

54. See e.g., contributions in Kooiman (1993).

55. The Norway co-management example and conclusions are in Hall-Arber and Finlayson (1997).

56. Scientists must present the results of their work in peer-reviewed journals for it to be accepted by their colleagues and the wider scientific world. That is the due process of science. On the other hand, scientists who speak out on issues beyond their expertise can lose credibility among peers. Nor will they necessarily find much credibility in the eyes of politicians, who will see their view as just another opinion. However, these things, although still widespread, are changing. As science becomes increasingly integrated, well-founded scientific arguments tend no longer to represent the view of a single leading biologist or chemist or physicist, but of a community whose various disciplines have fortified each other, and whose vigorous debate has illuminated, or at least located, areas of uncertainty. A competent scientist can thus present a viewpoint that integrates the social, economic and other (e.g., biological and physical) aspects of an issue. Whereas in the nineteenth century, individual scientists, naturalists then, felt qualified to speak or write about any scientific subject, twenty-first century scientists do this, with far more authority, based on the results of collaborative studies.

57. To go thus far is both mandatory and irresistible to responsible applied scientists (as has been put in a different way by the U.S. National Research Council "scientists must take the time to relate scientific knowledge to society in such a way that members of the public can make an informed decision about the relevance of research"). From the National Research Council's Committee on Science, Engineering, and Public Policy (CSEPP 1995).

58. This book, based directly on data assembled and analyzed since mid-1999 by the *Sea Around Us* Project team, was not intended to be a presentation and review of all the relevant work on the North Atlantic Ocean previously published by others (indeed, very few of the figures presented here are based on data external to the project). The result is that we did not cover all the issues relevant to fisheries impact on North Atlantic ecosystems; for example we did not deal with the impact of bottom trawling on bottom-living organisms and communities (Watling and Norse, 1998; Collie et al. 2000; Eleftheriou 2000; Hutching 2000). Also, we will have to elaborate elsewhere on the matters covered in our recent review of world fisheries (Pauly et al. 2002). We hope, however, that readers have appreciated the set of completely new concepts presented herein for envisioning fisheries impact on ecosystems, and new kinds of maps embodying these concepts. The *Sea Around Us* Project team will follow up on these, both by refining the analyses of the North Atlantic, and by expanding our coverage to include the Central and South Atlantic, and then the rest of the world's oceans. Please visit our website (www.saup.fisheries.ubc.ca) for updates and access to the catch and fish distribution database underlying the maps and other figures presented in this book.

References

ADB. 2001. *Poverty reduction: what's new and what's different?* Asian Development Bank, Manila.

ADBI and ADB. 2001. Asian cities in the 21st century. *Contemporary approaches to municipal management. Volume V. Fighting urban poverty.* Asian Development Bank Institute and Asian Development Bank, Manila.

Alder, J., T. J. Pitcher, D. Preikshot, K. Kaschner, and B. Ferriss. 2000. How good is good?: A rapid appraisal technique for evaluation of the sustainability status of fisheries in the North Atlantic, pp. 136–182. In: D. Pauly and T. J. Pitcher (eds.), Methods for evaluating the impact of fisheries on North Atlantic ecosystems. *Fisheries Centre Reports.* 8(2). [Available online at www.saup.fisheries.ubc.ca]

Alder, J., G. Lugten, R. Kay, and B. Ferriss. 2001. Compliance with international fisheries instruments for the North Atlantic, p. 55–80. In: T.J. Pitcher, U.R. Sumaila, and D. Pauly (eds.), Economic policy analysis of fisheries in the North Atlantic. *Fisheries Centre Research Reports.* 9(5). [Available online at www.saup.fisheries.ubc.ca]

Alverson, D.L., M.H. Freeberg, S.A. Murawski, and J.G. Pope 1994. A global assessment of fisheries bycatch and discards. *FAO Fish. Tech.* Pap. 339.

Anon. 2001a. Marine conservation. Net benefits. *The Economist,* 24 February, p. 83.

———. 2001b. Subsidies for sustainability–how Europe assists its fisheries sector. *Fishing Boat World* 12(12):9–11.

Anthea Brooks, L., and S.D. VanDeever (eds.). 1997. *Saving the seas: values, scientists, and international governance.* Maryland Sea Grant College, Maryland.

Arnason, R., R. Hannesson, and W.E. Schrank. 2000. Costs of fisheries management: the cases of Iceland, Norway and Newfoundland. *Mar. Policy* 24:233–243.

Bakun, A. 1996. *Patterns in the oceans: ocean processes and marine population dynamics.* University of California Sea Grant, San Diego and Centro de Investigaciones Biológicas de Noroeste, La Paz.

Bellwood, D.R., and T.P. Hughes. 2001. Regional-scale assembly rules and biodiversity of coral reefs. *Science* 292:1532–1534.

Bloom, D.E., P.H. Craig, and P.M. Maloney. 2001. *The quality of life in rural Asia.* Oxford University Press (China), Hong Kong.

Brothers, N.P., J. Cooper, and S. Løkkeborg. 1999. The incidental catch of seabirds by longline fisheries: worldwide review and technical guidelines for mitigation. FAO *Fish. Circular* 937.

Bundy, A., G.R. Lilly, and P.A. Shelton. 2000. A mass balance model of the Newfoundland-Labrador Shelf. *Can. Rep. Fish. Aquat. Sci.* 2310: xiv + 154 p.

Caddy, J.F., J. Csirke, S.M. Garcia, and R.J.R. Grainger. 1998. How pervasive is "fishing down marine food webs"? *Science* 282:1383–1383b.

Cadigan, S.T. 1999. Failed proposal for fisheries management and conservation in Newfoundland, 1855–1880, pp. 147–169. In: D. Newell and R.E. Ommer (eds.), *Fishing places, fishing people.* University of Toronto Press, Toronto.

Carson, R. 1951. *The Sea Around Us.* Oxford University Press, New York.

———. 1962. *Silent Spring.* Houghton Mifflin, New York.

Castillo, S., and J. Mendo. 1987. Estimation of unregistered Peruvian anchoveta (*Engraulis ringens*) in Peruvian catch statistics, 1951 to 1982, pp. 109–116. In: D. Pauly and I. Tuskayama (eds.), The Peruvian anchoveta and its upwelling ecosystem: Three decades of change. *ICLARM Studies and Reviews.* 15.

Charton, B., and J. Tietjen. 1989. Seas and oceans. *Collins Reference Dictionary.* Collins, Glasgow.

Christensen, V. 1995a. Ecosystem maturity—towards quantification. *Ecol. Modelling* 77:3–32

———. 1995b. A model of trophic interactions in the North Sea in 1981, the year of the stomach. *Dana* 11(1):1–28

———. 1996. Managing fisheries involving predator and prey species. *Rev. Fish Biol. and Fish.* 6:1–26.

———. 2001. Ecosystems of the past: how can we know since we weren't there?, pp. 26–34. In: S. Guénette, V. Christensen, and D. Pauly (eds.), Fisheries impacts on North Atlantic ecosystems: models and analyses. *Fisheries Centre Research Reports.* 9(4). [Available online at www.saup.fisheries.ubc.ca]

Christensen, V., and Pauly, D. 1998. Changes in models of aquatic ecosystems approaching carrying capacity. *Ecol. Applications* 8(1):104–109.

Christensen, V., and C. Walters. 2000. Ecopath with Ecosim: methods. capabilities and limitations, pp. 79–105. In: D. Pauly and T.J. Pitcher (eds.), Methods for evaluating the impacts of fisheries on North Atlantic ecosystems. *Fisheries Centre Research Reports.* 8(2). [Available online at www.saup.fisheries.ubc.ca]

Christensen, V., S. Guénette, H. Heymans, C. Walters, R. Watson, D. Zeller, and D. Pauly. 2001. Trends in the abundance of high-trophic level fishes in the North Atlantic, 1950 to 2000, pp. 1–25. In: S. Guénette, V. Christensen, and D. Pauly (eds.), Fisheries impacts on North Atlantic ecosystems: models and analyses. *Fisheries Centre Research Reports.* 9(4). [Available online at www.saup.fisheries.ubc.ca]

Christensen, V., S. Guénette, J.J. Heymans, C.J. Walters, R. Watson, D. Zeller, and D. Pauly. 2003. Hundred-year decline of North Atlantic predatory fishes. *Fish and Fisheries* (in press).

Chuenpagdee, R., and J. Alder. 2001. Sustainability ranking of the North Atlantic Fisheries, pp. 49–54. In: T.J. Pitcher, U.R. Sumaila, and D. Pauly (eds.), Economic policy analysis of fisheries in the North Atlantic. *Fisheries Centre Research Reports.* 9(5). [Available online at www.saup.fisheries.ubc.ca]

Cochrane, K. 2000. Reconciling sustainability, economic efficiency and equity in fisheries: the one that got away? *Fish and Fisheries* 1:3–21.

Coelho, M.L., and K.A. Stobberup. 2001. Portuguese catches of Atlantic cod (*Gadus morhua*) in Canada for the Period 1896–1969: Comparison with NAFO data, pp. 236–239. In: D. Zeller, R. Watson, and D. Pauly (eds.), Fisheries impacts on North Atlantic ecosystems: catch, effort and national/regional data sets. *Fisheries Centre Research Reports* 9(3). [Available online at www.saup.fisheries.ubc.ca]

Collie, J.S., G.A. Escanero, and P.C. Valentine. 2000. Photographic evaluation of the impacts of bottom fishing on benthic epifauna. *ICES J. Mar. Sci.* 57:987–1001.

Commission of the European Communities. 2000. Report from the Commission to the Council. *Preparation for a mid term review of the multi-annual guidance programmes* (MAGP). COM (2000) 272.

Coull, J.R. 1993. *World fisheries resources.* Routledge, London.

CSEPP. 1995. *On being a scientist. Responsible conduct in research.* Committee on Science, Engineering, and Public Policy. National Academy Press, Washington, DC.

Daan, N. 1997. Multispecies assessment issues for the North Sea, pp. 126–133. In: E.L. Pikitch, D.D. Huppert, and M.P. Sissenwine (eds.), *Global trends: fisheries management.* Amer. Fish. Soc. Symp. 20. American Fisheries Society, Bethesda, Maryland.

Dingsør, G. 2001. Norwegian un-mandated catches and effort. pp. 92–98. In: D. Zeller, R. Watson, and D. Pauly (eds.), Fisheries impacts on North Atlantic ecosystems: catch, effort and national/regional data sets. *Fisheries Centre Research Reports* 9(3). [Available online at www.saup.fisheries.ubc.ca]

Dobbs, D. 2000. *The great gulf: fishermen, scientists, and the struggle to revive the world's greatest fishery.* Island Press, Washington, DC.

Dyson, J.1977. *Business in great waters. The story of British fishermen.* Angus and Robertson, London.

Eleftheriou, A. 2000. Marine benthos dynamics: environmental and fisheries impacts. Introduction and overview. *ICES J. Mar. Sci.* 57:1299–1302.

Ehrlich, P.R. 1988. The loss of diversity: causes and consequences, pp. 21–27. In: E.O. Wilson (ed.), *Biodiversity*. National Academy Press, Washington, DC.

FAO. 1999a. Report of the FAO technical working group meeting on reduction of incidental catch of seabirds in longline fisheries, Tokyo, Japan, 25–27 March 1998. FAO *Fish. Rep.* (585).

————. 1999b. International plan of action for reducing incidental catch of seabirds in longline fisheries. International plan of action for the conservation and management of sharks. International plan of action for the management of fishing capacity. FAO, Rome.

————. 2000a. Advisory committee on fisheries research. Report of the working party on status and trends of fisheries, Rome, 30 November–3 December 1999. FAO *Fish. Rep.* (616).

————. 2000b. Report of the technical consultation on illegal, unreported and unregulated fishing, Rome, 2–6 October 2000. FAO *Fish. Rep.* (634).

————. 2001. Report of the joint FAO/IMO ad hoc working group on illegal, unreported and illegal fishing and related matters, Rome, 9–11 October 2000. FAO *Fish. Rep.* (637).

Fujiwara, M., and H. Caswell. 2001. Demography of the endangered North Atlantic right whale. *Nature*. 414:537–541.

Francis, D. 1991. *The great chase. A history of world whaling*. Penguin Books, Toronto.

Froese, R., and D. Pauly (eds.). 2000. FishBase 2000. *Concepts, design and data sources*. ICLARM, Los Baños, Philippines. 344 p. [Distributed with 4 CD-ROM; updates at www.fishbase.org]

Giberne, A. 1910. *The romance of the mighty deep*. Seeley and Co., Ltd., London.

Greer, J. 1995. *The big business takeover of U.S. fisheries: privatizing the oceans through individual transferable quotas*. Greenpeace, USA.

Guénette, S., T.J. Pitcher, and C. J. Walters. 2000. The potential of marine reserves for the management of northern cod in Newfoundland. *Bull. Mar. Sci.* 66(3):831–852.

Hall, S. J. 1999. *The effects of fishing on marine ecosystems and communities*. Blackwell Science, Oxford.

Hall-Arber, M., and A.C. Finlayson. 1997. Role of local institutions in groundfish policy, pp. 111–138. In: J. Boreman, B.S. Nakashima, J.A. Wilson, and R.L. Kendall (eds.), Northwest Atlantic groundfish: perspectives on a fishery collapse. *Amer. Fish. Soc.* Bethesda, Maryland.

Hardin, G. 1968. The tragedy of the commons. *Science* 162:1243–1248.

Heincke, F. 1913. Investigation on the plaice. General rapport I. Plaice fisher and protective regulation. Part I. *Rapp. P.-v. Réun. Counc. Int. Explor. Mer.* 17A:1–153.

Henderson, H. 1991. *Paradigms in progress: life beyond economics*. Knowledge Systems Inc., Indianapolis.

Hilton-Taylor, C., 2000. *IUCN red list of threatened species*. International Union for the Conservation of Nature (IUCN), Gland, Switzerland & Cambridge, UK.

Hinman, K. 2000. *Conservation in a fish-eat-fish world*. National Coalition for Marine Conservation, Leesburg, Virginia.

Holt, S.J. 1960. *Multilingual vocabulary and notation for fishery dynamics*. FAO, Rome.

Hoyt, E. 2001. *Whale Watching 2001: Worldwide tourism—numbers, expenditures and expanding socioeconomic benefits*. International Fund for Animal Welfare, Yarmouth Port, MA, USA.

Hutchings, P. 1990. Review of the effects of trawling on macrobenthic epifaunal communities. *Australian J. Mar. Freshwat. Res.* 41:111–120.

Hutchings, J.A. 2000. Collapse and recovery of marine fishes. *Nature* 406:882–885.

Independent World Commission on the Oceans. 1998. *The ocean our future*. Report of the Independent World Commission on the Oceans. Cambridge University Press, Cambridge, UK.

Jackson, J.B.C., M.X. Kirby, W.H. Berger, K.A. Bjorndal, L.W. Botsford, B.J. Bourque, R.H. Bradbury, R. Cooke, J. Erlandson, J.A. Estes, T.P. Hughes, S. Kidwell, C.B. Lange, G.S. Lenihan, J.M. Pandolfi, C.H. Peterson, R.S. Steneck, M.J. Tegner, and R.R. Warner. 2001. Historical overfishing and the recent collapse of coastal ecosystems. *Science* 293:629–637.

Jenness, D. 1972. *The Indians of Canada*. Sixth Edition. Natl. Mus. Canada Bull. (65).

Johnstone, K. 1977. *The aquatic explorers: a history of the Fisheries Research Board of Canada*. University of Toronto Press, Toronto.

Jones, G.P., M.J. Millcich, M.J. Emslie, and C. Lunow. 1999. Self-recruitment in a coral reef population. *Nature* 402:802–804.

Kaczynski, V.M., and D.L. Fluharty. 2002. European policies in West Africa: who benefits from fisheries agreements? *Marine Policy.* 26:75–93.

Kaschner, K., R. Watson, A.W. Trites, and D. Pauly. 2001. Estimating food consumption of marine mammals and trophic competition with fisheries in the North Atlantic, pp. 35–45. In: D. Zeller, R. Watson, and D. Pauly (eds.), Fisheries impacts on North Atlantic ecosystems: catch, effort and national/regional data sets. *Fisheries Centre Research Reports.* 9(3). [Available online at www.saup. fisheries.ubc.ca]

Kooiman, J. (ed.). 1993. *Modern governance: new government—society interactions.* SAGE, London, 280 p.

Kooiman, J., M. van Vliet, and S. Jentoft (eds.). 1999. *Creative governance: opportunities for fisheries in Europe*. Ashgate, Aldershot, UK.

Kurlansky, M. 1997. *Cod: A biography of the fish that changed the world*. Walker and Company, New York.

Larkin, P.A. 1977. An epitaph for the concept of maximum sustainable yield. *Trans. Amer. Fish. Soc.* 106:1–11.

Lee, M. (ed.). 2000. *Seafood lover's almanac*. National Audubon Society, Laval, Quebec.

Leopold, A. 1949. *A sand county almanac*. Oxford University Press.

Longhurst, A.R. 1995. Seasonal cycles of pelagic production and consumption. *Progress in Oceanography* 36:77–167.

———. 1998. *Ecological geography of the sea*. Academic Press, San Diego.

Longhurst, A.R., S. Sathyendranath, T. Platt, and C.M. Caverhill. 1995. An estimate of global primary production in the ocean from satellite radiometer data. *J. Plankton Res.* 17:1245–1271.

Mackinson, S. 2001. Representing trophic interactions in the North Sea in the 1880s, using the Ecopath mass-balance approach, pp. 35–98. In: S. Guénette, V. Christensen, and D. Pauly (eds.), Fisheries impacts on North Atlantic ecosystems: models and analyses. *Fisheries Centre Reports* 9(4). [Available online at www.saup.fisheries.ubc.ca]

Magnusson, K.G. 1993. Whaling: past, present and future, pp. 11–15. In: T.J. Pitcher and R. Chuenpagdee (eds.), Commercial whaling: the issues reconsidered. *Fisheries Centre Research Reports*. 1(1). [Available online at www.fisheries.ubc.ca]

Maguire, J.-J. 2001. Fisheries science and management in the North Atlantic, pp. 36–48. In: T.J. Pitcher, U.R. Sumaila, and D. Pauly (eds.), Fisheries impacts on North Atlantic ecosystems: evaluations and policy exploration. *Fisheries Centre Research Reports* 9(5). [Available online at www.saup.fisheries.ubc.ca]

McGlade, J. 1999. Bridging disciplines: the role of scientific advice, especially biological modelling, pp. 175–185. In: J. Kooiman, M. Van Vliet, and S. Jentoft, (eds.), *Creative governance: opportunities for fisheries in Europe*. Ashgate, Aldershot, UK.

McGoodwin, J.R. 1990. *Crisis in the world's fisheries*. Stanford University Press, Stanford.

Moloney, C. 1999. Approaches to integrating trophic modeling with physical oceanography, pp. 30–31. In: D. Pauly, V. Christensen, and L. Coelho (eds.), Proceedings of the EXPO '98 Conference on Ocean Food Webs and Economic Production, Lisbon, Portugal, 1–3 July 1998. ACP EU *Fish. Res. Rep.* (5).

Mowat, F. 1984. *Sea of slaughter*. McClelland and Stewart, Toronto.

Munro, G.R., and U.R. Sumaila. 2001. Subsidies and their potential impact on the management of the ecosystems of the North Atlantic, pp. 10–27. In: T.J. Pitcher, R. Sumaila, and D. Pauly (eds.), Economic policy analysis of fisheries in the North Atlantic. *Fisheries Centre Research Reports*. 9(5). [Available online at www.saup.fisheries.ubc.ca]

Murawski, S.A., R. Brown, H.-L Lai, P.J. Rago, and L. Hendrickson. 2000. Large-scale closed areas as a fishery-management tool in temperate marine systems: the Georges Bank experience. *Bull. Mar. Sci.* 66(3):775–798.

Myers, R.A., N.J. Barrowman, J.M. Hoenig, and Z. Qu. 1996a. The collapse of cod in eastern Canada: the evidence from tagging data. *ICES J. Mar. Sci.* 53:629–640.

Myers, R.A., J.A. Hutchings, and N.J. Barrowman. 1996b. Hypotheses for the decline of cod in the North Atlantic. *Marine Ecol. Progr. Ser.* 138:293–308.

————. 1997. Why do fish stocks collapse? The example of cod in Atlantic Canada. *Ecol. Applications* 7(1):91–106.

NRC. 1994. *An assessment of Atlantic bluefin tuna.* National Research Council, National Academy Press, Washington, DC.

————. 1999a. *Sharing the fish: toward a national policy on individual fishing quotas.* National Research Council, National Academy Press, Washington, DC.

————. 1999b. *Sustaining marine fisheries.* National Research Council, National Academy Press, Washington, DC.

NEAFC. 1985. *Handbook of basic texts.* North East Atlantic Fisheries Commission, Office of the Commission, London.

————. 1999. *Report of the seventeenth annual meeting, London, November* 1998. North East Atlantic Fisheries Commission, Office of the Commission, London.

Neutel, A.M., J.A.P. Heesteebeek, and P.C. de Ruiter. 2002. Stability in real food webs: weak links in long loops. *Science* 296:1120–1123.

OECD. 1997. *Toward sustainable fisheries.* OECD, Paris, 268 pp.

————. 2000. *Government financial transfers and resource sustainability: case studies.* Organisation for Economic Co-operation and Development, Directorate of Food, Agriculture and Fisheries, Paris.

Pauly, D. 1995. Anecdotes and the shifting baseline syndrome of fisheries. *Trends in Ecology and Evolution* 10(10):430.

————. 1996a. One hundred million tonnes of fish, and fisheries research. *Fish. Res.* 25(1):25–38.

————. 1996b. Biodiversity and the retrospective analysis of demersal trawl surveys: a programmatic approach, p. 16. In: Pauly, D. and P. Martosubroto (eds.), Baseline studies in biodiversity: the fish resources of western Indonesia. ICLARM *Studies and Reviews* 23.

Pauly, D., and V. Christensen. 1995. Primary production required to sustain global fisheries. *Nature* 374:255–257.

Pauly, D., V. Christensen, and C. Walters. 2000. Ecopath, Ecosim and Ecospace as tools for evaluating ecosystem impact of fisheries. ICES *Journal of Marine Science* 57:697–706.

Pauly, D., and M.L. Palomares. 2000. Approaches for dealing with three sources of bias when studying the fishing down marine food web phenomenon, pp. 61–66. In: F. Durand (ed.), Fishing down the Mediterranean food webs? Proceeding of a CIESM workshop, Kerkyra, Greece, 26–30 July 2000. CIESM *Workshop Ser.* 12.

————. 2001. Fishing down marine food webs: an update, pp. 47–56. In: L. Bendell-Young and P. Gallaugher (eds.), *Waters in peril.* Kluwer Academic Publishers, Dordrecht, The Netherlands.

Pauly, D., and T.J. Pitcher. 2000. Assessment and mitigation of fisheries impacts on marine ecosystems: a multidisciplinary approach for basin-scale inferences, applied to the North Atlantic, pp. 1–12. In: D. Pauly and T.J. Pitcher (eds.), Methods for evaluating the impacts of fisheries on North Atlantic ecosystems. *Fisheries Centre Research Reports.* 8(2). [Available online at www.saup.fisheries.ubc.ca]

Pauly, D., R. Froese, and V. Christensen. 1998a. Response to: how pervasive is "fishing down marine food webs"? *Science* 282:1383–1383b.

Pauly, D., T.J. Pitcher, and D. Preikshot (eds.). 1998b. Back to the future: reconstructing the Strait of Georgia ecosystem. *Fish. Centre Res. Rep.* 6(5). [Available online at www.fisheries.ubc.ca]

Pauly, D., V. Christensen, J. Dalsgaard, R. Froese, and F.C. Torres, Jr. 1998c. Fishing down marine food webs. *Science* 279:860–863.

Pauly, D., M.L. Palomares, R. Froese, P. Sa-a, M. Vakily, D. Preikshot, and S. Wallace. 2001. Fishing down Canadian aquatic food webs. *Can. J. Fish. Aquatic Sci.* 58:51–62.

Pauly, D., V. Christensen, R. Froese, A. Longhurst, T. Platt, S. Sathyendranath, K. Sherman, and R. Watson. 2000. Mapping fisheries onto marine ecosystems: a proposal for a consensus approach for regional, oceanic and global integrations, pp.13–22. In: D. Pauly and T.J. Pitcher (eds.), Methods for evaluating the impacts of fisheries on North Atlantic ecosystems. *Fisheries Centre Research Reports.* 8(2). [Available online at www.saup.fisheries.ubc.ca]

Pauly, D., V. Christensen, S. Guénette, T.J. Pitcher, U.R. Sumaila, C.J. Walters, R. Watson, and D. Zeller. 2002. Toward sustainability in world fisheries. *Nature* 418:689–695.

Pitcher, T.J. 2001a. Fisheries managed to rebuild ecosystems? Reconstructing the past to salvage the future. *Ecol. Applications* 11(2):601–617.

Pitcher, T.J. 2001b. FC and the MSC. *FishBytes,* the newsletter of the Fisheries Centre, University of British Columbia 7(2):1.

Pitcher, T.J., and D. Pauly. 1998. Rebuilding ecosystems, not sustainability, as the proper goal of fishery management, pp. 311–329. In: T. J. Pitcher, P. Hart, and D. Pauly (eds.), *Reinventing fisheries management.* Kluwer Academic Publishers, London.

Pitcher, T.J., and R. Watson. 2000. The basis for change 2: estimating total fishery extractions from marine ecosystems of the North Atlantic, pp. 40–53. In: D. Pauly and T.J. Pitcher (eds.), Methods for evaluating the impacts of fisheries on North Atlantic ecosystems. *Fisheries Centre Research Reports.* 8(2). [Available online at www.saup.fisheries.ubc.ca]

Pitcher, T.J., N. Haggan, D. Preikshot, and D. Pauly. 1999. "Back to the Future": a method employing ecosystem modeling to maximize the sustainable benefits from fisheries, pp. 447–466. In: Proceedings of the 16th Lowell Wakefield Fisheries Symposium. *Ecosystem approaches for fisheries management.* University of Alaska Sea Grant College Program.

Pitcher, T.J., M. Vasconcellos, S. Heymans, C. Brignall, and N. Haggan (eds.). 2002. Information supporting past and present ecosystem models of Northern British Columbia and the Newfoundland Shelf. *Fisheries Centre Research Report.* 10(1), 116 pp. [Available online at www.fisheries.ubc.ca]

Pope, J. 1989. Fisheries research and management in the North Sea: the next hundred years. *Dana* 8:33–43.

Porter, G. 2001. Fishing subsidies and overfishing: toward a structured discussion. Paper prepared for the UNEP Fisheries Workshop, Geneva, Switzerland.

Posadas, B.C., N.A. Ty, and E.B. Seraspe. 1986. Fish consumption in Iloilo: a consumer profile and behaviour study, pp. 179–210. In: D. Pauly, J. Saeger, and G. Silvestre (eds.), *Resources management and socioeconomics of Philippine marine fisheries*. University of the Philippines in the Visayas, College of Fisheries, Technical Report, Department of Marine Fisheries, No 10.

Rejwan, C., S. Booth, and D. Zeller. 2001. Unreported catches in the Barents Sea and adjacent waters for periods between 1950 and 1998, pp. 99–106. In: D. Zeller, R. Watson, and D. Pauly (eds.), Fisheries impacts on North Atlantic ecosystems: catch, effort and national/regional data sets. *Fisheries Centre Research Report*. 9(3). [Available online at www.saup.fisheries.ubc.ca]

Roach, C.M. 2000. Stewards of the Sea: a model for justice? pp. 67–82. In: H. Coward, R. Ommer, and T. Pitcher (eds.), *Just fish: ethics and Canadian marine fisheries*. Institute of Social Economic Research. Memorial University of Newfoundland. St. John's.

Roberts, C. 1999. Protected areas as a strategic tool for ecosystem management, pp. 37–43. In: D. Pauly, V. Christensen, and L. Coelho (eds.), Proceedings of the EXPO '98 conference on ocean food webs and economic production, Lisbon, Portugal, 1–3 July 1998. ACP EU *Fish. Res. Rep*. (5).

Roberts, C.M., J.A. Bohnsack, F. Gell, J.P. Hawkins, and R. Goodridge. 2001. Effects of marine reserves on adjacent fisheries. *Science* 294:1920–1923.

Russ, G.R., and A.C. Alcala. 1999. Management histories of Sumilon and Apo marine reserves, Philippines, and their influence on national marine resource policy. *Coral Reefs* 18:307–319.

Ruttan, L.M., F.C. Gayanilo Jr., U.R. Sumaila, and D. Pauly. 2000. Small versus large-scale fisheries: a multi-species, multi-fleet model for evaluating their interactions and potential benefits, pp. 64–78. In: D. Pauly and T.J. Pitcher (eds.), Methods for evaluating the impacts on North Atlantic ecosystems. *Fisheries Centre Research Reports*. 8(2). [Available online at www.saup.fisheries.ubc.ca]

Schrank, W.E., and B. Skoda. 2001. The cost of marine fishery management in eastern Canada: Newfoundland 1989/90 to 1997/98. In: R.S. Johnston and A.L. Shriver (eds.), Proceedings of the 10th International Conference of the International Institute of Fisheries Economics and Trade, July 10–14, 2000, Corvallis, Oregon. [CD-ROM; available from Dept. of Resource Economics, Oregon State University, Corvalis, OR.]

Sherman, K., and A.M. Duda. 1999. An ecosystem approach to global assessment and management of coastal waters. *Mar. Ecol. Progr.* Ser. 190:271–287. [updates in http://www.edc.uri.edu/lme/default.htm]

Sherman, K., N.A. Jaworski, and T.S. Smayda (eds.). 1996. The northeast shelf ecosystem. *Assessment, sustainability, and management*. Blackwell Science, Cambridge, MA.

Sobel, L.A. (ed.). 1975. *World food crisis*. Facts on File, Inc. New York.

Sumaila, U.R. 1999. Pricing down marine food webs. pp. 13–15. In: D. Pauly, V. Christensen, and L. Coelho (eds.), Proceedings of the '98 EXPO Conference on Ocean Food Webs and Economic Productivity, Lisbon, Portugal, 1–3 July 1998. ACP-EU *Fisheries Research Report*. 5.

————. 2001. Generational cost benefit analysis for evaluating marine ecosystem restoration, pp. 3–9. In: T.J. Pitcher, U.R. Sumaila, and D. Pauly (eds.), Economic policy analysis of fisheries in the North Atlantic. *Fish. Centre Res. Rep.* 9(5). [Available online at www.saup.fisheries.ubc.ca]

Sumaila, U.R., Y. Liu, and P. Tyedmers. 2001. Small versus large-scale fishing operations in the North Atlantic, pp. 28–35. In: T.J. Pitcher, U.R. Sumaila, and D. Pauly (eds.), Fisheries impacts on North Atlantic ecosystems: evaluations and policy explorations. *Fisheries Centre Research Reports* 9(5). [Available online at www.saup.fisheries.ubc.ca]

Swearer, S.E., J.E. Caselle, D.W. Lea, and R.R. Warner. 1999. Larval retention and recruitment in an island population of a coral-reef fish. *Nature* 402:799–802.

Tort, P. (ed.). 1996. *Dictionnaire du Darwinisme et de l'Evolution*. Presses Universitaires de France. Paris.

Trites, A.W., D. Pauly, and V. Christensen. 1997. Competition between fisheries and marine mammals for prey and primary production in the Pacific Ocean. *J. North West Atlantic Fish. Sci.* 22:173–187.

Tyedmers, P. 2001. Energy consumption of North Atlantic fisheries, pp. 12–34. In: D. Zeller, R. Watson, and D. Pauly (eds.), Fisheries impacts on North Atlantic ecosystems: catch, effort and national/regional data sets. *Fish. Centre Res. Rep.* 9(3). [Available online at www.saup.fisheries.ubc.ca]

Vasconcellos, M., S. Mackinson, K. Sloman, and D. Pauly. 1997. The stability of trophic mass-balance models: a comparative analysis. *Ecol. Modelling* 100:125–134.

Vitousek , P.M., P.R. Ehrlich, A.H. Ehrlich, and P.A. Matson. 1986. Human appropriation of the products of photosynthesis. *Bioscience* 36:368–373.

Wackernagel, M., and W.E. Rees. 1996. Energy efficiency comparison between the Washington and Japanese otter trawl fisheries of the Northeast Pacific. *Mar. Fish. Rev.* 37(4):21–24.

Walters, C.J. 1998. Designing fisheries management systems that do not depend upon accurate stock assessment, pp. 279–288. In: T. J. Pitcher, P. Hart, and D. Pauly (eds.), *Reinventing fisheries management*. Kluwer Academic Publishers, Dordrecht.

Walters, C. V. Christensen, and D. Pauly. 1997. Structuring dynamic models of exploited ecosystems from trophic mass-balance assessments. *Rev. Fish Biol. Fish.* 7(2):139–172.

Walters, C. D. Pauly, and V. Christensen 1998. Ecospace: prediction of mesoscale spatial patterns in trophic relationships of exploited ecosystems, with emphasis on the impacts of marine protected areas. *Ecosystems* 2:539–554.

Ward, B., and R. Dubos. 1972. *Only one earth. The care and maintenance of a small*

Ward, B., and R. Dubos. 1972. *Only one earth. The care and maintenance of a small planet.* W.W. Norton & Co., New York.

Warren, W. G., P.A. Shelton, and G. B. Stenson. 1997. Quantifying some of the major sources of uncertainty associated with estimates of harp seal prey consumption. Part I: Uncertainty in the estimates of harp seal population size. *J. Northw. Atlantic Fish. Sci.* 22:289–302.

Watson, R., S. Guénette, P. Fanning, and T.J. Pitcher. 2000. The basis of change: Part I. Reconstructing fisheries catch and catch and effort data, pp. 23–39. In: D. Pauly and T.J. Pitcher (eds.), Methods for evaluating the impacts of fisheries on North Atlantic ecosystems. *Fisheries Centre Research Reports.* 8(2). [Available online at www.saup.fisheries.ubc.ca]

Watson, R., and D. Pauly 2001. Systematic distortions of world fisheries catches. *Nature* 414:534–536.

Watson, R., A. Gelchu, and D. Pauly. 2001a. Mapping fisheries landings with emphasis on the North Atlantic, pp. 1–111. In: D. Zeller, R. Watson, and D. Pauly (eds.), Fisheries impacts on North Atlantic ecosystems: catch, effort and national/regional data sets. *Fisheries Centre Research Reports.* 9(3). [Available online at www.saup.fisheries.ubc.ca]

Watson, R., L. Pang, and D. Pauly 2001b. The marine fisheries of China: development and reported catches. *Fisheries Centre Research Reports* 9(1), 50 p. [Available online at www.saup.fisheries.ubc.ca]

Watling, L., and E.A. Norse. 1998. Disturbance of the seabed by mobile fishing gear: a comparison to forest clearcutting. *Conservation Biology.* 12(6):1180–1197.

Weber, M. 2002. *From abundance to scarcity: A history of U.S. marine fisheries policy.* Island Press, Washington, DC.

Went, A.E.J. 1972. Seventy years agrowing. A history of the International Council for the Exploration of the Sea. *Rapp. P. -v. Réun. Cons. Int. Explor. Mer.* 165.

Wilkinson, T. 2001. The seafood diet. *Wildlife Conservation* 104(1):22–29.

Wilson, E.O. 1998. *Consilience. The unity of knowledge.* A.A. Knopf, New York.

Yodzis, P. 2001. Must top predators be culled for the sake of fisheries? *Trends Ecol. & Evol.* 16(2):78–84.

de Young, B., R.M. Peterman, A.R. Dobell, E. Pinkerton, Y. Breton, A.T. Charles, M.J. Fogarty, G.R. Munro, and C. Taggart. 1999. Canadian marine fisheries in a changing and uncertain world. *Canadian Spec. Publ. Fish. Aquat. Sci.* 129.

Zarrilli, S., V. Jha, and R. Vossenair. 1997. *Eco-labelling and international trade.* McMillan Press, Hampshire, UK.

Zeller, D., and D. Pauly. 2000. How life history patterns and depth zone analysis can help fisheries policy, pp. 54–63. In: D. Pauly and T.J. Pitcher (eds.), Methods for evaluating the impacts of fisheries on North Atlantic ecosystems. *Fisheries Centre Research Report.* 8(2). [Available online at www.saup.fisheries.ubc.ca]

———. 2001. Visualisation of standardized life history patterns. *Fish and Fisheries.* 2(4):344–35.

Acronyms

CITES	Convention on International Trade in Endangered Species of Wild Fauna and Flora
CPI	consumer price index
EEZ	exclusive economic zone
FAO	Food and Agriculture Organization of the United Nations
ICCAT	International Commission for Conservation of Atlantic Tunas
ICES	International Council for the Exploration of the Sea
ICNAF	International Commission for the North West Atlantic Fisheries
ISOFISH	International Southern Oceans Longline Fisheries Information Clearing House
ITQ	individual transferable quota
IWC	International Whaling Commission

LME	large marine ecosystem
MPA	marine protected area (usually including a no-take marine reserve)
NAFO	Northwest Atlantic Fisheries Organization
NASCO	North Atlantic Salmon Conservation Organization
NEAFC	Northeast Atlantic Fisheries Commission
NGO	nongovernmental organization
OECD	Organization for Economic Co-operation and Development
SAUP	*Sea Around Us* Project
TAC	total allowable catch
UNCLOS	UN Convention on the Law of the Sea

Index